GREENFEAST

AUTUMN, WINTER

A BROWN VEGETABLE STOCK

Deep flavours. A herbal, umami-rich stock for winter cooking.

Makes about 2 litres

onions, medium 2
carrots 250g
celery 2 sticks
garlic a small, whole head
light miso paste 3 tablespoons
water 80ml, plus 3 litres

dried shiitake mushrooms 50g
rosemary 5 sprigs
thyme 10 sprigs
bay leaves 3
black peppercorns 12
dried kombu 10g

Set the oven at 180°C/Gas 4. Peel and roughly chop the onions, then place in a roasting tin together with the skins. Similarly chop the carrots and celery sticks, then mix with the onions and the head of garlic, separated into cloves.

Mix together the miso paste and 80ml of water, then stir into the vegetables, coating them lightly. Bake for about an hour, tossing the vegetables once or twice during cooking, until everything is brown, fragrant and toasty.

Transfer the roasted vegetables and aromatics to a deep saucepan, add the shiitake, rosemary, thyme, bay, peppercorns and the sheets of kombu, then pour a little of the reserved water into the roasting tin, scrape at the sticky, caramelised bits stuck to the tin, then pour into the saucepan. Add the remaining water. Bring to the boil, then lower the heat and leave to simmer, partially covered with a lid, for fifty minutes to an hour.

When you have a deep brown, richly coloured broth, tip through a sieve into a heatproof bowl or large jug and leave to cool. Refrigerate and use as necessary.

• Keeps for up to one week in the fridge.

stock, to bubble on the stove, to which you add pieces of hot toast, letting them slowly swell with the bosky, fungal, roasted flavours from the bowl.

WINTER PORRIDGE

A winter's day should start well. A steaming bowl of something to see us on our way. I invariably choose porridge. An oat-based slop to satisfy and strengthen, to bolster and soothe, to see me through till I get where I'm going. A sort of internal duffel coat. I doubt it will just be porridge of course, but porridge with bells and whistles: a trickle of treacle, a pool of crème fraîche, ribbons of maple syrup or a puddle of yoghurt. There may be golden sultanas and dried mulberries, pistachios or toasted almonds and perhaps some baked figs or slices of banana.

Porridge doesn't necessary mean oats. You could use rye grain or barley and milk or water as you wish. There might be salt or sugar, cinnamon or ground cardamom or toasted pumpkin seeds. If I remember, there will be stewed fruit too: apples perhaps, or dried apricots cooked with sugar or honey. Porridge is never just porridge in my house. It is a winter staple, one of the building blocks of the season and something I could never think of being without.

WINTER STOCK

A good vegetable stock is worth its weight in gold on a winter's day. As the nights draw in, we probably need a stock altogether deeper, richer and more ballsy than the delicate, vegetal liquids we might use in summer. Something that behaves more like a brown meat stock. Such a broth is immensely useful in my kitchen as a base for the heartier non-meat recipes that form the backbone of my daily eating, but also as something restoring to drink as you might a cup of miso. The colour must be dark and glossy, the flavour deeply, mysteriously herbal with a hint of mushroom and there should be a roasted back note, the sort you find in a long-simmered meat stock.

As you proceed, the kitchen will fill with the smell of onions, celery and carrots, which you roast with miso paste, then remove from the oven and simmer for a good hour with thyme, bay and shiitake. You could slip in a sheet of kombu for an extra layer of depth if you like.

The broth will need straining and separating from its spent aromatics, its deep, almost mahogany liquor dripping slowly into a glass bowl. The liquor can be used immediately, or kept in the fridge, covered, for up to a week.

Such a stock is a bowl of pure treasure. You can drink it like broth, dipping thick hunks of bread or focaccia into it; you can use it as base for a soup, adding steamed cauliflower or shredded cabbage, parsley and croutons, or add noodles, skeins of udon or little pasta stars to twinkle in the dark, mushroomy depths. Whenever the word 'stock' appears in a recipe, use it neat or let it down with a little water to taste. And it will freeze too, though I suggest in small containers, so it defrosts quickly.

And when all is said and done, is there anything quite so restoring as coming home to a bowl of deeply layered, smoky

WINTER
BASICS

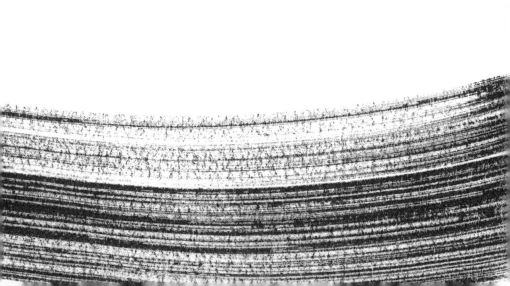

The idea that so many people's everyday eating is going through such a change and that meat is no longer our first thought when working out what we want to eat, is heartening to say the least. Apparently I am not the only person for whom meat is still looked forward to but as a once- or twice-a-week treat, not the knee-jerk star of every meal. I knew this was happening (you would have to live under a stone not to) but I genuinely hadn't realised how widely and quickly the change has come about.

Yes, vegan cooking and full-blown vegetarianism is on the rise, but there are far more people who seem to prefer a less rigid approach to their eating. This makes sense on so many levels, but when all is said and eaten it is simply that the options for cooking without meat have never been more varied or delicious. There has never been a better time to celebrate the move towards a mostly plant-based diet.

This is also the season for 'something on toast'. By toast I mean not only bread cut thick and rough-edged, but toasted bagels and crumpets, muffins and naan. Any soft dough that will crisp under the grill and will support a cargo of vegetables or is happy to be slathered with a thick wave of crème fraîche or hummus, roast vegetables or perhaps cheese to melt and bubble.

And, of course, there must be pudding. An early autumn crumble of damsons and almonds. Chocolate puddings (you really must make the ones with dulce de leche). Ginger cake with a cardamom cream and a custard pudding set with cake and apples. There will be nut-encrusted shortbreads with blood orange and baked apples with crisp crumbs and cranberries. I expect syllabubs and baked pears on the table, pastries laden with a golden dice of apples and scones pebble-dashed with nibs of dark chocolate.

I probably eat more puddings during the cold months, but mainly at the weekend. The main course recipes in this book are predominantly for two; the puddings, though, are all for four or more. You can't really make a tart for two or a tiny batch of scones. The recipes are made for sharing with friends and family. That said, most of them are rather fine eaten the following day. Especially those little chocolate puddings.

A NOTE ON VOLUME 1. *Greenfeast: spring, summer.*

Like all my books, the first volume of *Greenfeast* was written from and about my own kitchen. That it found itself welcomed by quite so many came as something of a pleasant surprise. I have lost count of the number of people who in the last few months have told me that this is the way they eat now, as an 'almost vegetarian.'

My cold weather eating is more substantial than the food I eat for the rest of the year. Dinner becomes about one main dish rather than several lighter ones, and the focus shifts towards keeping warm. On returning home I will now happily spend an hour cooking. Maybe a little longer.

The oven gets more use at this time of year, the grill and griddle probably less. More food will come to the table in deep casseroles and pie dishes. I dig out my capacious ladle for a creamed celeriac soup as soft as velvet. The temperature of the plates and bowls will change. We want to hold things that warm our hands, a sign of the happiness to come.

There will be carbs. They protect and energise us. They bring balm to our jagged nerves. (Winter is nature's way of making us eat carbohydrates.) Crusts – of pastry, breadcrumbs and crumble – add substance; potatoes fill and satisfy and there is once again a huge sourdough loaf on the table. Rice and noodles are no longer a side dish, and now become the heart and soul of dinner.

My autumn and winter cooking is every bit as plant-based as the food I make in the summer; it just has a bit more heft to it. Shallow bowls of rice cooked with milk and thyme in the style of a risotto. A verdant, filling soup of Brussels sprouts and blue cheese. A saffron-coloured stew of sour cream, herbs and noodles. Translucent fritters in a pool of melted cheese. Golden mush-rooms astride a cloud of soft, yellow polenta. There is a tangle of noodles and tomato, peppery with chilli; roast parsnips and baked pumpkin; a wide earthenware dish of sweet potatoes and lentils glowing like a lantern, a herb-freckled crumble of leeks and tomato or swedes and thyme in a pastry crust and a tarte Tatin of soft golden shallots and autumn apples. It is all here, between these pumpkin-coloured covers.

INTRODUCTION

Dinner is different in winter. The change starts late on a summer's evening, when you first notice the soft, familiar scent of distant woodsmoke in the sudden chill of the evening air. Then, a day or two later, a damp, mushroomy mist hovers over the gardens and parks. Later, you notice the leaves have turned silently from yellow ochre to walnut. Autumn is here once again. You may sigh, rejoice or open a bottle. For many, this is the end of their year. For me, this is when it starts, when warmth, and bonhomie come to the fore. Energy returns.

With the change of weather, supper takes on a more significant role. We are suddenly hungry. Once the nights draw in, I am no longer satisfied by plates of milky burrata and slices of sweet, apricot-fleshed melon. No more am I content with a bowl of couscous with peaches, soft cheese and herbs for dinner. What I crave now is food that is both cosseting and warming, substantial and deeply satisfying. Food that nourishes but also sets me up for going back out in the cold and wet. And yet, I still find my diet is heavily plant-based with less emphasis on meat. It is simply the way it has progressed over the years and shows little sign of abating.

At the start of the longest half of the year, our appetite is pricked by the sudden drop in temperature, and as evenings get longer, we have the opportunity to spend a little more time in the kitchen. To mash beans into buttery clouds. Simmer vegetable stews to serve with bowls of couscous. To bring dishes of sweet potato to melting tenderness in spiced cream. And of course, the pasta jar comes out again.

Greenfeast   xiii

xii   Greenfeast

Contents

Tom Kemp

Tom Kemp has had a couple of careers. He trained formally as a theoretical computer scientist and followed a sequence of post-doctoral research and programming posts. In parallel, he was learning to be an artist by studying the writings of ancient manuscripts, not their content but how they were made. In particular, he worked out the details of a Roman signwriting technique which has informed all his brushwork, both readable and abstract. This calligraphic training led to a deeper pursuit of writing in general and artworks in many media, including graffiti and digital work. Along the way he learned to make porcelain vessels on a potter's wheel, an activity he describes as 'calligraphy in 3D', and these now form the large surfaces on which he continues to write.

tomkemp.com
Instagram @tom_kemp_

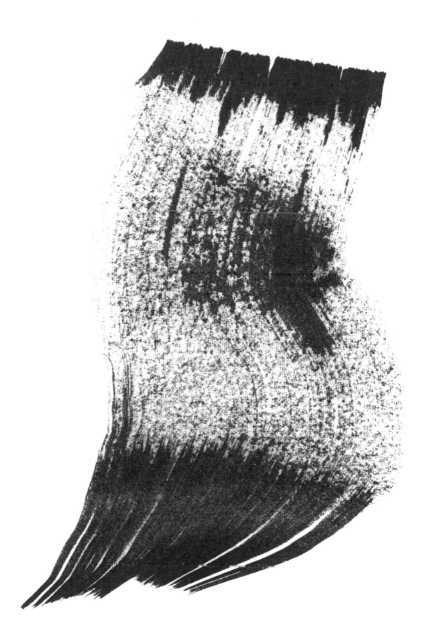

For James

4th Estate
An imprint of HarperCollins*Publishers*
1 London Bridge Street
London SE1 9GF
www.4thEstate.co.uk

First published in Great Britain by 4th Estate in 2019

5 7 9 8 6 4

Text copyright © Nigel Slater 2019

All recipe photographs © Jonathan Lovekin 2019
Except p. 183 and p. 279 © Nigel Slater 2019

Brushstrokes copyright © Tom Kemp 2019

A catalogue record for this book is available from the British Library

ISBN 978-0-00-821377-0

Design by David Pearson

Typeset by GS Typesetting

Printed in Neografia, Slovakia

MIX
Paper from
responsible sources
FSC
www.fsc.org   FSC C007454

This book is produced from independently certified FSC paper to ensure
responsible forest management.

Find out more about HarperCollins and the environment at
www.harpercollins.co.uk/green

GREENFEAST

AUTUMN, WINTER

Nigel Slater

Photography by
Jonathan Lovekin

4th ESTATE • *London*

Also by Nigel Slater:

Greenfeast: spring, summer
The Christmas Chronicles
The Kitchen Diaries III: A Year of Good Eating
Eat
The Kitchen Diaries II
Tender, Volumes I and II
Eating for England
The Kitchen Diaries I
Toast – the story of a boy's hunger
Appetite
Real Food
Real Cooking
The 30-Minute Cook
Real Fast Food

Nigel Slater is an award-winning author, journalist and television presenter. He has been food columnist for the *Observer* for over twenty-five years. His collection of bestselling books includes the classics *Appetite* and *The Kitchen Diaries* and the critically acclaimed two-volume *Tender*. He has made cookery programmes and documentaries for BBC1, BBC2 and BBC4. His memoir *Toast – the story of a boy's hunger* won six major awards and is now a film and stage production. His writing has won the James Beard Award, the National Book Award, the Glenfiddich Trophy, the André Simon Memorial Prize and the British Biography of the Year. He lives in London.

nigelslater.com
Twitter @nigelslater
Instagram @nigelslater

OATS, DRIED MULBERRIES, DATE SYRUP

The solace of porridge. The sweetness of dried fruits.

Serves 2

porridge oats 100g	cream or crème fraîche
dried mulberries 50g	4 tablespoons
golden sultanas 75g	date syrup 2 tablespoons

Put the oats and 400ml of water into a small saucepan and bring them to the boil. Add a good pinch of salt and stir the oats continuously for four or five minutes with a wooden spoon until the porridge is thick and creamy.

Divide between two bowls, then add the dried mulberries and golden sultanas. Add spoonfuls of crème fraîche, then trickle over the date syrup.

• I often use dried apricots in place of the mulberries, but cranberries and dried cherries are good alternatives.
• The sweetness of the date syrup can be balanced by a spoon or two of stewed tart apples.

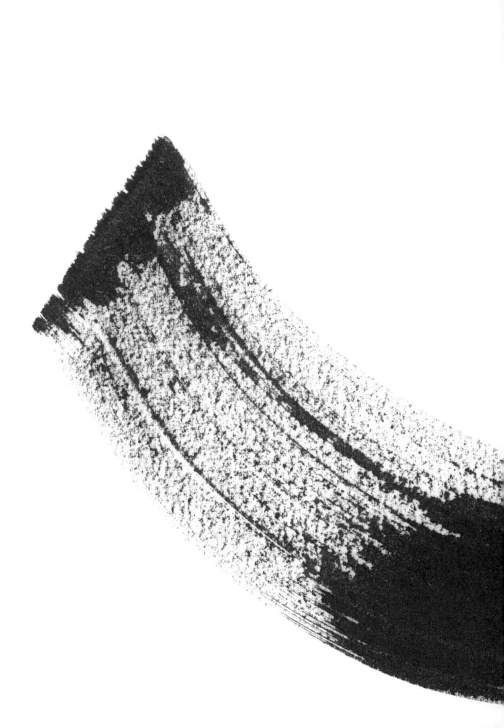

IN A PAN

On a winter's evening, I warm butter and olive oil in a shallow-sided, cast-iron pan, turn the heat down low and use it to fry thin slices of potato, Jerusalem artichoke or fennel. Vegetables that will crisp or soften as you wish, and to which I can add other ingredients at will – sweet black grapes and whole parsley leaves to the crisp artichokes; peas and salty cheese to the softened fennel. I do the same with slices of pumpkin or butternut squash, then introduce feta or breadcrumbs or perhaps a fried egg whose yolk will double as an impromptu sauce.

A heavy frying pan in which you can leave things to cook at a moderate temperature is worth its weight in gold. It is the gift I would give to a kid leaving home. The possibilities are endless. Mushrooms, sliced and sautéed with herbs to pile on a mound of silken hummus; beans whose outsides slowly crisp in olive oil and are then tossed with tomatoes and a wobbly egg of burrata; Brussels sprouts fried with miso paste to a deep walnut brown, then forked through sticky brown rice. They all give a substantial green and deeply savoury supper. The list is endless.

The success often lies in the pan itself. A pitted or wobbly-based pan will produce uneven results. Sometimes you need a thin-bottomed pan in which to flash-fry, other times a pan as heavy as possible that will hold the heat and which can be left to do its task while you get on with other elements of dinner. Choose your weapon.

It is worth finding a suitable lid. Especially if you are cooking vegetables that need to be brought gently to tenderness before a final crisping, such as potatoes, parsnips and carrots. The sort of heavyweight pans I find so useful for slow winter cooking often come without a lid, so it is not a bad idea to find one that fits before you leave the shop.

I am very fond of my old iron sauté pans, but they do need a bit of care when you first get them home. A good oiling with linseed oil, a long, slow bake in the oven and a careful dry before putting them away will give them a chance to develop a patina, a naturally non-stick layer that will, unlike a commercial non-stick finish, see you through a lifetime of suppers.

ARTICHOKES, BEANS, GREEN OLIVES

Crisp beans and fried artichokes. Dinner from the deli.

Serves 2

green olives, stoned 200g
olive oil 100ml
basil 20 leaves
lemon juice 75ml
parsley leaves from a small bunch
black garlic 2 cloves
olive oil, for frying 2 tablespoons

haricot beans 1 × 400g can
fine ground polenta 6 tablespoons
eggs 2
artichokes in oil 350g
groundnut oil, for deep frying

Put the olives into the jug of a blender, then add the olive oil, basil leaves, lemon juice, parsley leaves and black garlic. Reduce to a thick purée.

Warm the olive oil in a shallow pan that doesn't stick, drain the beans and add them to the pan, then let them cook over a moderate heat, turning occasionally, until they are lightly crisp.

Put the polenta on a large plate. Break the eggs into a shallow dish and mix the yolks and whites together lightly with a fork. If the artichokes are whole, then slice them in half. Roll each half in the beaten egg, then place in the polenta and turn over, pressing down firmly to coat evenly.

Heat the groundnut oil, add the artichokes and fry till golden and crisp. Lift each from the oil and drain briefly on kitchen paper before dividing between two plates. Serve with the crisp beans and the olive paste and, if you wish, a wedge of lemon.

• Make a black olive paste if you prefer, using stoned black olives. I would also be tempted to add a pinch of dried chilli flakes when you blend the ingredients.
• You can use breadcrumbs instead of polenta. Sieve fresh dry crumbs on to a plate and roll the artichokes in them after coating in egg.

AUBERGINES, GINGER, TAMARIND

Hot, sweet, sour.

Serves 2

lime juice 100ml (about 2 limes)
ginger 30g
fish sauce 3 teaspoons
palm sugar 4 teaspoons
hot red chilli 1
hot green chilli 1
tamarind paste 4 teaspoons
groundnut oil 2 tablespoons

aubergines 300g

For the apple yoghurt:
mint leaves 12
a small apple
white wine vinegar 2 tablespoons
natural yoghurt 200ml

Put the lime juice into a mixing bowl. Peel and grate the ginger, stirring the resulting paste into the lime juice. Pour in the fish sauce, then stir in the palm sugar until dissolved.

Finely chop the red and green chillies, removing the seeds if you wish, then add them to the marinade with the tamarind paste and groundnut oil, combining the ingredients thoroughly.

Cut the aubergines in half lengthways, then into wedges as you might a melon. Now cut each wedge in half. Put the aubergines into the marinade, turn to coat and set aside for a good half hour. During this time they will soften a little.

Make the apple yoghurt: finely chop the mint leaves and put them in a small mixing bowl. Grate the apple into the bowl, it can be as coarse or as fine as you wish, then stir in the white wine vinegar and yoghurt, cover and set aside.

To cook the aubergines, heat a cast-iron griddle (and switch on the extractor). Place the aubergines on the griddle and leave to brown on the underside. Turn, loosening them from the griddle with a palette knife, and

(continued)

brown the other side. Keep the heat low to moderate, and make sure they are cooked right through – they must be fully tender.

Serve the aubergines, hot from the griddle, with the apple yoghurt sauce.

• Cut the aubergines in slices or wedges, as the fancy takes you. There will, I assure you, be much smoke, so switch on the extractor or open a window. Better still, cook them outside on the barbecue. Arm yourself with a palette knife to gently prise them from the bars of the griddle. I like to keep the heat no hotter than medium to give the aubergine time to cook through to the middle, a process you can speed up by covering the aubergines with an upturned metal bowl (or a lid, if your griddle has one). If you prefer, rather than the sour apple dressing, make a dressing of olive oil, lime juice and coriander leaves.

• A twist of noodles, tossed with the merest splash of sesame oil, could be a suitable accompaniment here, as would long-grain rice, steamed and seasoned with black pepper and sesame seeds.

BRUSSELS SPROUTS, BROWN RICE, MISO

The savour of miso. The homeliness of brown rice.

Serves 2–3

brown sushi rice 190g

Brussels sprouts 750g

groundnut oil 2 tablespoons

light miso paste 1 tablespoon

Japanese pickles (tsukemono)

2 tablespoons

Put the rice in a bowl, cover with warm water, then run your fingers through the grains. Drain, repeat, then tip into a saucepan, cover with 5cm of cold water and set aside for half an hour.

Wash and trim the sprouts, then cut each in half. Bring the soaked rice to the boil in its soaking water, add half a teaspoon of salt, cover and lower the heat so the water simmers. Leave for thirty minutes or until the rice is approaching tenderness. Remove the pan from the heat and leave to rest for ten minutes before removing the lid.

Warm the oil in a shallow pan. Toss the sprouts with the miso paste, then transfer to the hot oil, moving them round the pan as they become crisp and pale golden brown.

Remove the lid from the rice, run a fork through the grains to separate them, then divide between two or three bowls. Spoon the miso sprouts into the rice and add some of the Japanese pickles.

• Fried in a little oil, the miso paste forms a fine crust on the outside of the sprouts. Serve them as an accompaniment if you wish, but I like them as the star of the show, tucked into a bowl of sticky rice and scattered with salty Japanese pickles. I serve this as it is, but also as a side dish for slices of cold roast pork and its crackling. This is sticky rice, my favourite, but you don't want it in lumps, so running the tines of a fork through the cooked grain is a good idea.

BURRATA, BEANS, TOMATOES

Milky snow-white cheese. Toasted beans. Peppery basil.

Serves 2

garlic 3 cloves	cherry tomatoes 250g
olive oil	basil leaves a handful
cannellini beans 1 × 400g can	burrata 2 × 250g balls

Flatten the garlic cloves with the blade of a kitchen knife, then peel away the skins. Warm four tablespoons of olive oil in a shallow pan and add the garlic, letting it cook briefly over a moderate heat. Drain the cannellini beans.

Cut the tomatoes in half, pour a little more oil into the pan, then add the tomatoes and the drained cannellini. Fry briefly, for four or five minutes, until the beans are starting to crisp a little.

Tear the basil leaves and add to the beans, stirring them in gently, until they start to wilt. Divide the beans and tomatoes between two plates, add the burrata and trickle with olive oil.

• The beans will crisp deliciously around the edges if you leave them to fry in the hot oil. Stirring them too often will cause them to break up as they develop their golden shell.

• Cannellini beans are my first choice here, but butter beans are worth considering too. Green flageolet don't seem to work quite as well, though I am not entirely sure why.

• This is one of the lighter recipes in this volume, yet each time I make it, I am surprised by how satisfying it is.

BUTTERNUT, BREADCRUMBS, CURRY POWDER

Sweet golden squash. Warm, spicy curry. Crisp crumbs.

Serves 2

onions, medium 2	ground turmeric 1 teaspoon
carrots, large 300g	vegetable stock 500ml
groundnut oil 3 tablespoons	panko breadcrumbs 6 tablespoons
butternut squash 500g	parsley, chopped 4 tablespoons
curry powder 2 teaspoons	togarashi 1–2 teaspoons

Peel and roughly chop the onions and carrots, then put them in a large saucepan with the oil and place over a moderate heat. Let the vegetables cook for ten to fifteen minutes until the onions are pale gold.

Slice the butternut into 2cm-thick rounds, deseed and peel it if you wish. When the onions are nicely golden, stir in the curry powder, ground turmeric and a little salt and fry briefly, then pour in the vegetable stock and bring to the boil. Tuck in the slices of squash and lower the heat to a simmer. Leave for fifteen minutes, then remove the squash and place half of the sauce in a blender. Process to a smooth purée, then return to the pan and keep at a low bubble for five minutes.

Toast the panko crumbs in a dry pan till golden, then toss with the chopped parsley and togarashi. Slide the squash back into the sauce for a couple of minutes, sprinkle with the parsley crumbs, then serve.

• The sauce is based on a classic katsu, which flatters the sweetness of the butternut. You can turn up the heat if you wish with a little more togarashi seasoning or even a splash of chilli sauce.
• The warmly spiced sauce is also worth trying with baked aubergine or roasted parsnips.
• Togarashi, the Japanese spice mix, can be found in major supermarkets, Japanese food shops and online.

BUTTERNUT, FETA, EGGS

Crisp, light, sweet, salty.

Makes 9 fritters. Serves 3

butternut squash 300g
garlic 2 cloves
groundnut oil
eggs 2
feta cheese 200g

plain flour 4 tablespoons
thyme leaves, chopped 1 tablespoon
parsley, chopped 3 heaped tablespoons
groundnut or vegetable oil,
   for deep frying

Peel and remove the seeds from the butternut squash. Push the squash through a spiraliser to give long, thin strings.

Peel the garlic and thinly slice it. Warm a little groundnut oil in a large, shallow pan, then add the garlic, let it sizzle for a couple of minutes then, as it starts to colour, drop in the squash and fry for five or six minutes till the colours are bright and the squash is tender but far from falling apart.

Separate the eggs. Make a batter by mixing together the egg yolks, crumbled feta cheese, plain flour, a grinding of black pepper and the chopped thyme leaves and parsley. Beat the egg whites till frothy, then fold into the batter. Toss the threads of butternut with the batter.

Warm enough groundnut oil in a deep, heavy pan to fry the fritters. When the oil is at 180°C, take a large spoonful of the batter-coated butternut and lower into the hot oil. Repeat with a further three or four, frying for three or four minutes till crisp and golden in colour. As each fritter is ready, remove with a draining spoon and rest on kitchen paper. Continue with spoonfuls of the batter until you have nine fritters. Serve hot.

• So good are these little fritters that I have tried them with other vegetables too, including shredded courgette (a success) and beetroot (less so). It is worth having something to dip them into, such as a cucumber, mint and yoghurt dip or a bowl of especially creamy hummus.

EGGS, EDAMAME, BEAN SPROUTS

A soft pillow of egg. A tangle of vegetables.

Serves 2

edamame beans, podded 200g

spring onions 8

pak choi 200g

garlic 3 cloves

large green chillies 2

groundnut oil 4 tablespoons

bean sprouts 200g

eggs 6

nigella seeds 2 teaspoons

coriander a handful

Bring a pan of water to the boil, add the edamame and boil till tender – about eight minutes. Drain and refresh in a bowl of iced water.

Finely chop the spring onions, discarding the roots and any tough dark green leaves. Shred the pak choi. Peel and thinly slice the garlic. Finely slice the chillies.

Warm half the groundnut oil in a large, shallow pan, fry the spring onions, garlic and chillies till soft, then add the shredded pak choi and lastly the bean sprouts, tossing them in the hot oil and cooking for three or four minutes till softened.

Break the eggs into a bowl and beat them lightly with a fork. Add the cooked and drained edamame, the fried vegetables and aromatics. Season with a little sea salt and black pepper and fold in the nigella seeds and coriander.

Warm the remaining oil in a large metal-handled frying pan, pour in the omelette mixture and fry over a moderate heat for about eight minutes, until the edges have set and the middle is still almost liquid. Heat the oven grill. Place the frying pan under the grill and continue cooking for two or three minutes until the centre of the omelette is lightly set. (Ideally, it should be a little *baveuse*, verging on the point of setting.) Cut in half and serve. *(continued)*

• To the basic mixture you can add pretty much any vegetable you have to hand, from fried mushrooms to steamed shredded cabbage. The cooking time is brief, so most vegetables will have to be lightly cooked first. Brassicas such as long-stemmed sprouting broccoli work very well, as do any late autumn beans. I especially like steamed mustard greens.

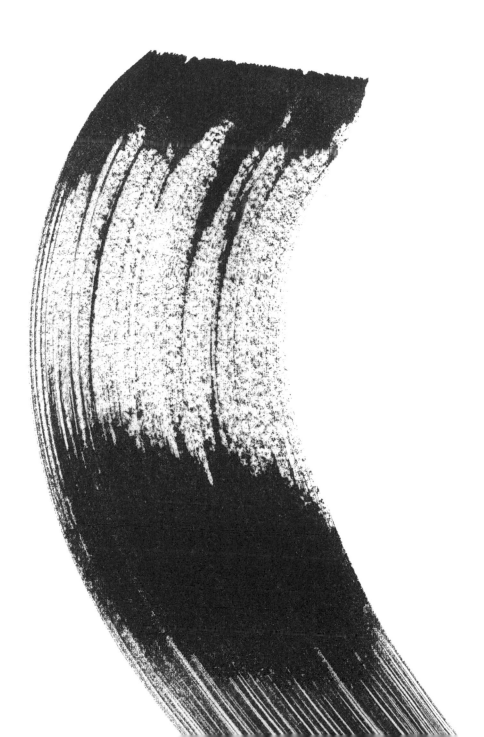

FENNEL, PEAS, HALLOUMI

Fresh green flavours for a golden autumn day.

Serves 2

fennel 300g
olive oil 3 tablespoons
halloumi 250g

For the dressing:

frozen peas 250g
basil leaves 30g
mint leaves 20g
olive oil 150ml

Trim and thinly slice the fennel, no thicker than a pound coin. Warm the 3 tablespoons of oil in your largest frying pan, place the fennel in the pan in a single layer, then season lightly and cook until the fennel is light gold on the underside. Turn each piece over and continue cooking until soft and translucent.

Slice the halloumi into pieces slightly thicker than the fennel and place them in the pan, tucking them in wherever there is a gap, overlapping where there isn't. Let the cheese toast and turn patchily gold.

Put the frozen peas in a colander and run them under the cold tap for a few minutes till they have defrosted. Let them drain. Whizz the peas, basil and mint leaves and the oil in a food processor till almost smooth – a slightly rough texture is good – then spoon over the cheese and fennel and serve. If you have a few fennel fronds, add them at the very end.

JERUSALEM ARTICHOKES, PISTACHIOS, GRAPES

Golden carbs. Black grapes.

Serves 2–3

a lemon

Jerusalem artichokes 200g

shallots, small 6

black grapes 150g

olive oil 3 tablespoons

golden sultanas 4 tablespoons

shelled pistachios 4 tablespoons

parsley leaves a handful

Halve the lemon and squeeze the juice into a bowl. Peel the Jerusalem artichokes, then cut into rounds approximately 0.5cm thick. Put the artichokes into the bowl with the lemon juice and toss together. This will prevent them going brown.

Peel the shallots and cut them in half from stem to root. Halve and deseed the grapes. Warm the olive oil in your largest shallow pan. Add the artichokes and shallots to the hot oil, turning them over when their underside is golden and they are starting to soften.

As the artichokes become tender, add the grapes and sultanas and lastly, the pistachios and parsley.

• Small potatoes can successfully replace the artichokes. You could also use parsnips. I find it best to use parsnips that have been steamed for ten minutes before frying. They are more tender that way.

• Should you find it easier, you can roast the artichokes instead of frying them, adding the remaining ingredients to the pan once the artichokes are golden.

MUSHROOMS, HUMMUS, HERBS

Creamed chickpeas. Sizzling mushrooms.

Serves 2

double cream 250ml

chickpeas 1 × 400g can

coriander leaves and stems 10g

dill 15g

parsley 10g

king oyster mushrooms 400g

olive oil 4 tablespoons

butter 30g

garlic 4 cloves

Warm the cream and drained chickpeas in a medium-sized saucepan for five minutes over a moderate heat. Tip them into a blender or food processor with the coriander, dill and parsley and process to a thick cream. Transfer back to the saucepan and season.

Slice the mushrooms into 1cm-thick pieces. Warm the olive oil and butter in a shallow pan, add the slices of mushroom and cook till golden on both sides. You may need to add a little more oil. As each mushroom browns, remove to a plate or piece of kitchen paper.

Peel and thinly slice the garlic, cook briefly in the pan until it turns gold, then return all the mushrooms to the pan. Warm the hummus, spoon onto plates, then add the sizzling mushrooms and garlic.

• Mushrooms, fried with butter and garlic, are just one possibility here; the soft, herb-speckled purée lends itself to sitting under deep-fried artichokes or roast parsnips, baked tomatoes or wedges of baked cabbage.

• Such purées can be made and kept in a small bowl overnight in the fridge, their surface splashed with olive oil and tightly covered. Stored like this, their texture will thicken and they can be spread on slices of hot toast for a quick bite to eat when you arrive home, hungry.

PARSLEY, PARMESAN, EGGS

Warm, soft, parsley-freckled drop scones. (Picture overleaf.)

Makes 6

parsley leaves 15g
self-raising flour 180g
baking powder 1 teaspoon
a large egg

milk 220ml
Parmesan, grated 5 heaped
  tablespoons
butter a little

Roughly chop the parsley leaves. Put the flour in a large mixing bowl, add the baking powder and combine. (You can sieve the two together if you wish.) Break the egg into a bowl and beat lightly with a fork to combine white and yolk, mix in the milk, then stir into the flour.

Add the grated Parmesan and chopped parsley to the batter. Melt the butter in a small saucepan, then remove from the heat. In a medium, non-stick or well-used frying pan, pour a couple of tablespoons of the melted butter and let it warm over a moderate heat. Pour in a sixth of the batter, making a round approximately the size of a digestive biscuit. Repeat with two more, then let them cook for four or five minutes, checking the underside regularly for colour.

When they are puffed and golden, use a palette knife to carefully turn each one over. Leave for a further three or four minutes, then lift out and keep warm. A sound test for doneness is to touch the centre of each scone with your finger. It should feel lightly springy. Continue with the remaining batter. Serve with the pumpkin hash overleaf.

• The drop scones can be flavoured with chopped thyme or rosemary, basil or tarragon. In which case I would serve them with grilled tomatoes, lightly crushed with a fork, or chopped spinach softened with a little cream.

PUMPKIN, ONIONS, ROSEMARY

Sweet and sticky squash.

Enough for 6

pumpkin or butternut squash 700g	butter 30g
onions, medium 2	olive oil 2 tablespoons
	rosemary 3 sprigs

Peel the pumpkin and cut the flesh into cubes roughly 3 x 3cm. Peel and roughly chop the onions. Warm the butter and olive oil in a large, shallow pan, add the pumpkin and onions and let them cook, with a regular stir, for about ten minutes. Finely chop the rosemary leaves, discarding the stalks, add to the pan with a little sea salt and coarsely ground black pepper, then cover with a lid and leave to cook over a low to moderate heat for about fifteen minutes.

Check the onion and pumpkin occasionally to make sure they aren't browning too much. They are done when soft and easily crushed between your fingers. Serve with the drop scones on the previous page.

• As well as a side dish for the drop scones this can be used as the stuffing for an omelette or frittata, or served as a vegetable dish piled onto steamed rice.
• Use the recipe above with courgettes instead of pumpkin.

RAINBOW CHARD, EGG, NOODLES

A joyful tangle of noodles and greens.

Serves 2

a red chilli, medium
a green chilli, medium
a bunch of coriander 25g
garlic 1 clove
ginger a 15g piece
rainbow chard 100g

eggs 3
groundnut oil 5 tablespoons
fresh udon noodles
soy sauce to taste
sesame oil a dash

Thinly slice the chillies. Cut the coriander stems into small pieces the length of a matchstick and reserve the leaves. Peel and thinly slice the garlic. Peel the ginger, then cut into skinny matchsticks. Remove the leaves from the chard and roughly shred them, then chop the stems into short pieces.

Break the eggs into a small bowl and beat them. Warm half the oil in a large, shallow pan over a moderate heat, pour in the beaten egg and leave to set, checking the underside after a minute or two. When the omelette is golden, flip over and let the other side cook for a minute or two, then remove from the pan and tear the omelette into small pieces.

Wipe the pan clean with kitchen paper, add the remaining oil and let it sizzle. Add the ginger, garlic and chillies to the pan and fry till golden and fragrant. Add the coriander and chard stems and continue cooking for two minutes, then add the soft noodles and toss everything together.

Add in the soy sauce and a little sesame oil, add the chard and coriander leaves and the torn omelette, and continue cooking for a minute till all is hot and sizzling.

TOFU, RADISH, PONZU

A gossamer-thin, crisp crust. Trembling tofu. A salty citrus dressing.

Serves 2

soft tofu 340g
ponzu sauce 4 tablespoons
sesame oil 3 teaspoons
rice vinegar 3 tablespoons
spring onion 1

radishes 4
coriander leaves a handful
cornflour or potato flour 6 tablespoons
ginger a thumb-sized lump
oil, for deep frying

Drain the tofu on kitchen paper. Mix together the ponzu, sesame oil and rice vinegar. Trim and finely chop the spring onion, then thinly slice the radish and add both to the dressing together with the coriander leaves.

Put the cornflour or potato flour into a shallow bowl or deep plate. Cut the tofu into six large cubes. Finely grate the ginger to a purée.

Warm the oil to 180°C in a deep pan. Gently toss the cubes of tofu in the flour, then lower into the oil and fry for three or four minutes till light gold in colour. Divide the dressing between two deep bowls, lift the tofu from the oil with a draining spoon, then lower three pieces into each bowl, top with the puréed ginger and eat while the tofu is still hot and crisp.

ON TOAST

Two little words. So much joy. I sometimes think 'on toast' might be the best two words in the world. That slice of toast could be heavy with butter, shining pools forming on its surface; saturated with olive oil, glistening bright with crystals of sea salt, or perhaps its golden crust is shimmering with Marmite or honey.

Toast can be a round of 'white sliced' popping up from the toaster; thick air-pocketed sourdough, its crust caught black under the grill; neat rectangles of dark rye, all treacle and aniseed. Then again it could be a muffin, ripped rather than sliced in half, or a crumpet whose buttery cargo has trickled deep into its holes. Toast can be focaccia or ciabatta, fruit loaf or panettone, naan or roti. In short, any dough whose surface will toast crisply and hold something delicious.

I often tuck a piece of thick toast underneath a wedge of roasted pumpkin or aubergine, the bread soaking up the juices as it cooks. Almost anything, from broccoli you have cooked in butter on the hob to a stew whose juices deserve a thick pillow to soak them up, can be piled on top of toast.

It is difficult to think of anything more instantly satisfying than a thick wedge of golden bread under a pool of melted cheese. But that is just a start. I like to cook slices of leek or onion in butter till silky, then hide them underneath a layer of Cheddar or Wensleydale; mix grated beetroot and apple with goat's curd or soft cheese and sit it on sourdough or rye.

We can extend the very meaning of toast to include any piece of dough whose surface has been deliciously crisped by the heat. Many is the time I have commandeered everything from warm oatcakes to a white sliced loaf to use as a supporting role for cooked vegetables or melted cheese. Even a crumpet fits the bill, especially when spread with a layer of cream cheese and topped with lightly cooked spinach. And I could never forget using naan as a soft cushion on which to put torn pieces of mozzarella with tomatoes or an English muffin spread thick with hummus and fried mushrooms.

There are no rules really. The toast should be thick enough to support whatever we put on top of it and it should be hot and freshly made. Other than that, we are surely free to mix and match toasts and toppings at will. Authenticity is of little interest here; what matters is that the two are compatible.

A slice of toasted bread has the ability to make anything more substantial. A lump of cool, milky burrata with a trickle of peppery olive oil; a trio of roasted tomatoes, their skins blackened from the oven and scented with basil and butter; a spoon or three of last night's stew, brought to a steaming simmer, or the bread, cut thick and toasted and dunked into a bowl of soup where it will slowly sponge up every bit of goodness and savour.

BEETROOT, APPLE, GOAT'S CURD

Hot toast, cool curd, crisp seeds.

Serves 2

pickled beetroot 100g
a small, sharp apple
cider vinegar 2 tablespoons
sunflower seeds 2 tablespoons
poppy seeds 2 tablespoons

rye or sourdough bread
 2 thick slices
goat's curd 200g
dill, chopped 2 tablespoons

Coarsely grate the beetroot into a medium-sized mixing bowl. Grate the apple into the beetroot but don't mix them together. Sprinkle the cider vinegar over the apple and beetroot.

In a dry pan, toast the sunflower seeds for three or four minutes till golden and fragrant, then mix them with the poppy seeds. Toast the bread lightly on both sides.

Stir the goat's curd into the apple and beetroot, together with a little salt and half the seeds and chopped dill. Spread the apple and goat's curd onto the toasts in thick waves, scatter over the remaining seeds and dill and eat immediately, while the toast is still hot.

• A light stir is all you need to mix the curd with the apple and beetroot. Over-mixing will result in a rather pink covering for your toast.
• Instead of the goat's curd, try fromage frais or a fresh, fluffy goat's cheese.

CRUMPETS, CREAM CHEESE, SPINACH

The comfort of crumpets. The goodness of greens.

Serves 2

spinach 100g	spring onion 1
tarragon, chopped 1 tablespoon	cream cheese 200g
chives, chopped 1 tablespoon	crumpets 4
parsley, chopped 2 tablespoons	

Wash the spinach, remove any thick stems, then pile the still-wet leaves into a saucepan. Place the pan over a moderate heat, cover tightly with a lid, then let the leaves steam for a minute or two. Lift the lid and turn the leaves, so they soften evenly. When they are bright green and glossy, remove from the pan, squeeze the leaves almost dry and chop roughly.

Mix the chopped tarragon, chives and parsley together. Finely slice the spring onion and mix with the herbs, then stir into the cream cheese. Season with black pepper and a little salt. Toast the crumpets till golden, then spread each generously with some of the herb cream cheese. Pile some of the spinach leaves on top and tuck in.

• The crumpet's holes have the ability to soak up anything we spread it with, from butter to herbed cream cheese. Rather than sit on top of the toasted crust, the melted cheese, cream, Marmite or whatever trickles down through the holes to become part of the crumpet itself. An utterly joyous conception and the reason for keeping a packet in the freezer.

• Once a crumpet is toasted, I find it best to cover the surface generously with something delicious, then return it briefly to the grill. Crumpets don't hold the heat well, so an extra minute or so under the heat will ensure they are piping hot. *(continued)*

• A few delicious cheese-based ideas for your crumpets: butter and grated farmhouse cheeses; slices of soft Brie-type cheese and sharp fruit chutney such as apple or apricot; a spread made from equal amounts of butter and grated Parmesan; cream cheese flecked with dill and chopped pickled cucumber; cream into which you have melted Taleggio and added a few thyme leaves; crème fraîche and black cherry jam.

EGGS, SPINACH, BREAD

Spiced, eggy toasts.

Serves 2

spinach 200g	dried chilli flakes 1 teaspoon
curry powder 1 teaspoon	eggs 4
ground turmeric 1 teaspoon	ciabatta 4 thick slices
ground cumin ½ teaspoon	butter 40g

Wash the spinach leaves, then put them, still very wet, into a deep saucepan for which you have a tight-fitting lid. Place over a high heat and let the spinach steam, turning it from time to time, until the leaves are dark green and wilted. A matter of five minutes or less. Remove the spinach from the pan, squeeze the leaves to remove most of the water, then chop finely.

Put the curry powder, turmeric, cumin and chilli flakes in a dry, shallow pan and toast over a moderate heat for a minute or two till fragrant. Tip them into a medium-sized mixing bowl, then break the eggs into the bowl, add a half teaspoon of salt, stir in the spinach leaves and beat lightly to combine whites, yolks, spinach and toasted spices.

Dunk the slices of ciabatta in the spiced batter, making sure they are well saturated. Melt the butter in a shallow pan, either non-stick or one with a good patina. As the butter warms, lower the bread into the pan, moderating the heat so it browns lightly on the base, making sure the butter doesn't burn – a matter of three or four minutes. Turn with a palette knife and brown the other side. Drain lightly on kitchen paper before eating.

• A substantial snack more than a meal, but a useful recipe to have to hand. Use any soft bread, preferably of the airy and light sort. I have known these be made for breakfast on a Sunday morning, an occasion to which they seem most fitting. *(continued)*

• Use your favourite curry powder or own masala. You may have your own recipe or maybe a commercial brand, but either way add more or less to suit your taste.

LEEKS, CAERPHILLY, MUSTARD

The sweetness of leeks in butter. The quiet heat of mustard.

Serves 2

a large leek
butter 30g
plain flour 1 heaped tablespoon
double cream 250ml

Caerphilly 100g
grain mustard 1 tablespoon
Dijon mustard 1 tablespoon
crumpets 4

Thinly slice the leek, discarding any particularly thick, dark green leaves, and wash thoroughly. Put the still-wet leek in a deep saucepan over a low heat with the butter and cover with a lid. Leave to cook, with the occasional stir, for ten to fifteen minutes until they are softened. Don't let the leeks brown.

Stir the flour into the leeks, continue cooking for a couple of minutes, then pour in the double cream. Coarsely grate the Caerphilly into the leeks, then season with the grain and smooth mustards, and a little salt and black pepper.

Heat an overhead grill. Toast the crumpets. Spoon the leek and cheese mixture over them and return to the grill. Cook until the surface is patchily golden. (I like mine blistered dark brown here and there.)

• Onions make a sound substitute for the leeks. Onions take longer to cook, so fry them over a low to moderate heat, stirred regularly for a good thirty to forty minutes until they are deep golden brown.
• You can use a more strongly-flavoured cheese if you wish, even a blue variety such as Shropshire Blue or Stichelton, in which case I would forgo the mustards and add a spoonful more cream.

MUSHROOMS, BLUE CHEESE, TOASTING MUFFINS

Toasted dough. Golden fungi. A savoury butter.

Serves 2

chestnut mushrooms, small 150g
soft blue cheese such as
   Dolcelatte 50g
butter 30g
garlic 2 cloves

olive oil 4 tablespoons
thyme leaves 2 teaspoons
the juice of half a lemon
toasting muffins 2

Thinly slice the mushrooms. Cream together the blue cheese and butter. Peel and thinly slice the garlic.

Warm the olive oil in a shallow pan, add the garlic and let it soften. Add the mushrooms to the pan, followed by the thyme, and fry for six or seven minutes till they and the garlic are golden. Grind over a little black pepper and squeeze the lemon juice into the mushrooms.

Split the muffins horizontally, toast them, then spread with the blue cheese butter. Pile the mushrooms on top and eat.

NAAN, MOZZARELLA, TOMATOES

Crisp dough. Soft cheese. Sweet-sharp tomatoes.

Makes 2

mozzarella 125g	spring onions 3
feta 100g	naan 2
cherry tomatoes 150g	coriander leaves a handful

Set the oven at 220°C/Gas 8.

Coarsely grate the mozzarella into a bowl, then crumble the feta into it. Cut the tomatoes into quarters and roughly slice the spring onions. Mix everything together and grind in a little black pepper.

Place the naan on a baking sheet and divide the tomato, mozzarella and feta mixture between them. Bake for twelve minutes, or until the naan is crisp and the cheese is pale gold in patches. Scatter with coriander leaves.

• Naan provides a generously-sized base for anything you might usually put on a pizza. More interesting than most commercially-made pizza bases, the best naan are lighter and fluffier and will crisp splendidly in the oven with their cargo of melting cheese.

• I love the cultural mix here of warm Indian dough with an Italian topping. Mozzarella provides a pool of molten cheese, to which you can add something more strongly flavoured, such as Parmesan or Pecorino, or a mixture of cheeses you wish to use up.

IN THE OVEN

A black, wet, winter's night. The kitchen is dark save for the light from the oven. A light as sweet as treacle. In the oven is a wedge of pumpkin or a humble baked potato, a roasting cauliflower or golden roots (parsnip, carrot, sweet potato) whose edges are slowly browning. There may be the scent of warm thyme or garlic. The burnt-sugar-and-butter smell of roasting onions or the resinous notes of rosemary twigs baking. Your oven may be home to a wide earthenware dish of sweet potato and cream, its edges bubbling and caramelising in the heat. Or it could just be witness to a single baked tattie, snowflakes of sea salt on its crisping skin, its insides slowly turning to a floury butter-yellow fluff.

My oven gets twice the use in winter it does during the other seasons. Occasionally I will slide a potato or more often a sweet potato on the bottom shelf and leave it to bake whilst I make a salad – of watercress and fennel or red treviso and oakleaf lettuce. Winter salad leaves are crisp, bitter and pale (chicory, Castelfranco, frisée), and benefit from citrus notes of orange and grapefruit, of toasted nuts warm from the pan, and from dressing spiked with mustard, grated horseradish or ginger.

I like the thought that whilst baking or roasting no flavour has the chance to escape, every bit of it trapped by the walls of the oven. In reality, a crust forms on top of the food rather than underneath, as it does with food cooked on the hob. A crust that seems to seal in flavour. But these of course are the romantic notions of a hungry cook, not those of the scientist. I know which I would rather be.

There is an ease to food that is baked or roasted. The oven does much of the work for us. There is no moving food around the pan, turning or stirring. No tinkering with the heat. That said, it is often a good idea to baste something as it cooks – a pumpkin spooned with the butter that has melted around it; a mushroom that is baking in a pool of olive oil and spices will benefit from a little TLC.

The dish in which you bake or roast matters little as long as it is strong enough not to buckle in the heat. You can bake in an aluminium tray if you have to. In my experience, the thinner the tin, the more likely the contents are to burn. Deep-sided dishes are good for baking, less so for roasting. Vegetables cooked in a deep-sided dish are less likely to toast and brown, as they tend to produce steam. To get a good toasted crust all round, roast in a shallow-sided tin.

The one oven-based supper to which I return time and again is that of simple roast vegetables. Peeled parsnips or sliced aubergines, onions or red peppers that you have tossed in olive oil and butter, salt, pepper and herbs and roast at a high temperature till the edges are sweet and dark. I let them cool, then dress them with their caramelised cooking juices and a squeeze of lemon or some red wine vinegar. I eat them on their own or with a mash of canned beans or a ladleful of pale golden polenta, or trickle them with a dressing of tahini and yoghurt. Opening the fridge on a winter's night and finding a dish of yesterday's roast vegetables is as good as an impromptu supper gets. For which read: make twice as much as you need, leaving some to come home to tomorrow.

ARTICHOKES, WINTER ROOTS, SMOKED SALT

Deep drifts of mash. Toasted, smoky vegetables.

Serves 4

For the mash:
celeriac 750g
Jerusalem artichokes 250g
butter 50g
hot vegetable or chicken stock
    250ml

For the roast roots:
carrots, small to medium 8

parsnips 2
garlic 6 cloves
beetroots, small raw 4
thyme 8 sprigs
butter 75g
sunflower seeds 3 tablespoons
parsley, chopped 2 heaped
    tablespoons
smoked salt 2 teaspoons

Set the oven at 200°C/Gas 6. Peel the celeriac and cut into large pieces, about the size of roast potatoes. Put them in a roasting tin. Peel the artichokes, then add them to the roasting tin. Break the butter into small pieces and dot over the vegetables, season, then roast for about an hour until lightly browned. They mustn't crisp like roast potatoes, but instead develop pale gold, lightly caramelised edges. Test them for softness; a skewer should slide through them with ease.

Scrub or peel the carrots as you wish, then cut them in half lengthways. Do the same with the parsnips, peeling them if their skins are tough (as they often can be) and cutting them in half. Put the carrots and parsnips in a roasting tin. Tuck the garlic cloves, still in their skins, amongst the carrots. Trim the little beetroots, cut into quarters, then add to the tin along with the sprigs of thyme and the butter. Season generously, then bake for about an hour. Turn the vegetables over once or twice as they roast.

Tip the roast celeriac and artichokes into a food processor, add most of the stock and reduce to a smooth purée, introducing more of the stock as

(continued)

necessary. Check the seasoning, adding more salt and pepper as you wish. In a small pan, toast the sunflower seeds over a high heat till they smell nutty, then mix with the chopped parsley and smoked salt.

Spoon the mash on to a serving dish, add the roast vegetables, scatter with the seeds and parsley and serve.

In the oven   71

BRUSSELS SPROUTS, SMOKED MOZZARELLA, DILL

Greens, herbs and cream.

Serves 3

Brussels sprouts 300g
butter 40g
olive oil 2 tablespoons
dill, chopped 15g
smoked mozzarella 250g
double cream 250ml

For the crumbs:
butter 40g
coarse breadcrumbs a handful
dill, chopped 3 tablespoons

Set the oven at 200°C/Gas 6. Trim and shred the Brussels sprouts. Warm the butter and olive oil in a shallow pan, then add the sprouts and cook for two minutes until they brighten in colour. Fold in the chopped dill, a little salt and a grind or two of black pepper. Cut the smoked mozzarella into thick slices.

Put half the sprouts into a baking dish, place most of the mozzarella amongst them, then a second layer of sprouts, finishing with the remaining cheese.

Make the crumbs: warm the butter in a shallow pan, add the crumbs and cook till golden, then stir in the chopped dill. Pour the cream over the sprouts and cheese, scatter with the crumbs and dill and bake for twenty-five minutes or until bubbling.

• A dish to warm the soul. You could use shredded cabbage, cauliflower or lightly cooked kale in place of the sprouts. The mozzarella could be replaced by a blue cheese of some sort, which would flatter the greens.
• Should you have any leftover cooked pasta, you could incorporate it with the sprouts and cheese.

BUTTER BEANS, PEPPERS, AUBERGINES

Clouds of mash. Sweet, sharp juices.

Serves 4

small red peppers 400g	rosemary 3 large sprigs
cherry tomatoes 250g	olive oil 5 tablespoons
small aubergines 400g	butter beans 1 × 650g jar or
garlic 6 fat cloves	2 × 400g cans

Set the oven at 200°C/Gas 6. Place the whole peppers, cherry tomatoes and aubergines in a roasting tin. Tuck the garlic cloves, still in their skins, and the rosemary among the vegetables, then spoon over the olive oil. Let the vegetables roast in the preheated oven for thirty to forty minutes, until they are approaching softness, then push them to one side of the roasting tin. Drain the butter beans and tip them into the tin. Stir the beans to coat them in oil and roasting juices, then return to the oven and cook for a further twenty minutes until all is soft and golden.

Remove the garlic from the roasting tin and squeeze each clove from its skin into the bowl of a food processor. Tip in the warm beans and process to a thick, fluffy purée. Check the seasoning, then pile the purée on to a serving plate and place the vegetables on top. Spoon over any juices.

• If you can track down the large, flat Judión beans in jars then I recommend them for this. They are particularly soft and buttery and produce a cloud-like mash. Canned butter beans work well enough, are cheap as chips and easier to locate. You may like to add a thick slice of butter as you blend them. The vegetables can be roasted ahead of time and served warm or even cold, but the beans must be blended while they are still warm, if you are to achieve a smooth finish. *(continued)*

• Once the weather cools and summer slides towards autumn, you could roast slices of orange-fleshed squash instead of the aubergines or use parsnips in place of the butter beans. The real point of the dish is the sweet, herb-infused roasting juices trickling into the cloud of mash.

CABBAGE, BERBERE SPICE, CRUMBS

Toasted cabbage. Crisp crumbs. Green peas.

Serves 2

cabbage 300g
olive oil 12 tablespoons
fresh white breadcrumbs 80g

Berbere spice mix 1 teaspoon
frozen peas 100g
mint leaves 12

Set the oven at 200°C/Gas 6. Slice the cabbage into 2cm-thick pieces and put them on a foil-lined baking sheet. Trickle over six tablespoons of the oil and bake for twenty-five minutes.

Warm three tablespoons of the oil in a shallow pan, add the crumbs and toast them till golden, tossing them from time to time. Stir in the Berbere spice mix.

In a food processor blend the peas with the remaining oil and the mint leaves until lightly chopped. Stir into the warm crumbs and let the peas thaw in the residual heat. Remove the cabbage from the oven, slide onto plates with a fish slice or palette knife, then scatter with the peas and crumbs.

• A crisp, rock hard-white cabbage is probably more suitable for roasting than an open-leaved dark green Savoy, which tends to cook to a crisp too easily.

• This recipe works with Brussels sprouts too. Slice them thickly and lay them out on the baking sheet. The result will be messier to serve, but will taste just as good.

• Berbere spice mixes are available from any major supermarket or speciality grocers. The mix is sweet and warm rather than hot and usually contains chilli, cumin, ginger, cardamom, nutmeg and fenugreek. Good for adding to melted butter and tossing with steamed long-grain rice or couscous.

CARROTS, SPICES, PANEER

Sweet vegetables. Earthy spice.

Serves 2–4

small carrots, assorted colours
    and sizes 800g
groundnut oil 5 tablespoons
coriander seeds 1 teaspoon

cumin seeds 1 teaspoon
nigella seeds 1 teaspoon
black mustard seeds 1 teaspoon
paneer 250g

Set the oven at 200°C/Gas 6. Trim and scrub the carrots and put them in a large roasting tin. Pour over the oil and toss gently to coat. Roast for forty to forty-five minutes till tender, tossing them halfway through.

Warm the coriander, cumin, nigella and mustard seeds in a dry pan for a couple of minutes. As soon as the spices smell toasted and are starting to pop, crush them coarsely using a pestle and mortar, or give them a brief ride in a spice mill. Scatter the spices over the carrots with a little salt and toss gently, then remove the carrots from their tin and transfer to a serving dish.

Place the tin over a moderate heat and add the paneer, crumbling it as you go. Let the paneer fry, with the occasional turn, until it is golden. Scatter the paneer over the carrots and serve.

• If paneer isn't your thing, it is well worth making this with feta cheese or halloumi instead. I have also brought these roast, spiced vegetables out with soft, fresh goat's cheese to add in cloud-like mounds at the table.
• Parsnips, cut into thin lengths, can work here too and take this particular spicing well.

CAULIFLOWER, ONIONS, BAY

Cloves, spring onions and bay. Flavours to soothe.

Serves 4

a cauliflower, about 1kg in weight
olive oil 6 tablespoons
white onions, large 2
butter 75g
milk 500ml

bay leaves 3
cloves 3
black peppercorns 8
spring onions 4

Set the oven at 200°C/Gas 6. Trim the cauliflower, removing any imperfect or tough leaves, keeping any small, young ones in place. Put the cauliflower stem down in a roasting tin, pour five tablespoons of the olive oil over and season with sea salt and ground black pepper. When the oven is up to heat, roast the cauliflower for fifty minutes to an hour, or until the florets are golden brown. Check its tenderness with a metal skewer – it should go through the thickest part of the cauliflower with ease.

While the cauliflower roasts, make the sauce. Peel and roughly chop the onions. Warm the remaining olive oil and the butter in a deep pan over a gentle heat, add the onions and let them cook, covered by a lid, till soft and translucent. Let them take their time, they need a good twenty-five minutes, and stir them regularly, so they do not colour.

Pour the milk over the softened onions, add the bay leaves, cloves and peppercorns and bring to the boil. Remove the pan from the heat and cover, leaving the milk to infuse for forty minutes.

Remove the bay, peppercorns and cloves, then use a stick blender (or, better I think, put the onions and milk into a blender) and reduce to a smooth and creamy sauce. Season carefully, then finely chop the spring onions. Return the onion sauce to its pan, stir in the spring onions and warm over a moderate heat, stirring almost continuously, till hot. Lift the cauliflower from the oven and transfer to a serving dish. Pour over the onion sauce and serve.

CHEDDAR, TARRAGON, EGGS

Light, airy. A cheese and tarragon pudding.

Serves 2

To prepare the dish:	butter 55g
a knob of butter	plain flour 50g
Parmesan, finely grated	eggs, large 4, separated
2 tablespoons	strongly-flavoured, firm cheese, such
	as Yarg, Gruyère, Cheddar, grated
For the pudding:	or cut into small pieces 100g
milk 300ml	tarragon leaves, chopped
a bay leaf	2 tablespoons
a small onion	Parmesan, grated 1 tablespoon

Lightly butter the inside of a deep soufflé or similar dish, measuring 20cm across the top, then scatter in one tablespoon of the grated Parmesan, tipping the dish from side to side to make sure the cheese sticks to the butter.

Put a baking sheet in the oven and set at 200°C/Gas 6. Bring the milk to the boil in a small pan together with the bay leaf and peeled onion. Turn off the heat and let it sit for a few minutes.

Melt the butter in a small, heavy-based saucepan, stir in the flour and leave over the heat for two or three minutes, stirring almost continuously. Stir in the warm milk, gradually at first (you can chuck the bay and onion at this point – they have done their work), then a little faster, till you have a thick sauce. Let it come to an enthusiastic bubble, then lower the heat and let the mixture simmer for at least five minutes, until it thickens.

If your sauce looks lumpy, whisk it fiercely until it is smooth. Remove from the heat, cool briefly, then stir or whisk in the egg yolks, one at a time. Work quickly, otherwise the egg will cook before it gets stirred in.

Stir in the grated cheese and the chopped tarragon. In a large bowl and using a large balloon whisk, beat the egg whites till stiff and frothy. Fold

(continued)

them gently but firmly into the cheese sauce, then immediately scrape into the buttered dish.

Smooth the top lightly, scatter with the remaining Parmesan, then place on top of the hot baking sheet and bake for twenty-five to thirty minutes.

To test if your pudding is done, push the dish firmly with your oven glove: it should shudder but not wobble violently. The crust should be pale to mid-brown, the centre should be soft and oozing. Remove and serve immediately. If you stick your spoon in and the middle is too liquid, put it back in the oven. It will still rise.

• The cheese is up to you. This can be a good way of using up bits you have lurking in the fridge. However, the flavour is best when the cheese has a good, strong character. Waxy, budget cheese doesn't have the clout – a strong farmhouse cheese with some bite to it is what you need.

CHEESE, THYME, GRAPES

Molten, oozing cheese. Cold sweet grapes.

Serves 2

a whole small Camembert or
   Tunworth cheese, preferably
   one in a wooden box
thyme 4 sprigs

black peppercorns 6
olive oil 2 tablespoons
crusty bread, to serve
grapes, to serve

Set the oven at 200°C/Gas 6. Remove the cheese from its waxed paper and return it to the little wooden box. Wrap the underside and edges of the box tightly with foil to prevent the cheese leaking. Put the thyme sprigs and peppercorns on the surface of the cheese and moisten them with the olive oil.

Bake for twenty minutes, or until the cheese is molten inside. Serve with bread, preferably with a good crisp crust, and some sweet black grapes.

• Check that the wooden box the cheese comes in is stapled, not glued. If glued then it is inclined to come undone in the oven and your liquid cheese will escape.
• I sometimes add half a dozen juniper berries, lightly crushed, to the surface of the cheese as well.

CHICKPEAS, RADICCHIO, BUTTER BEANS

Soft, floury beans and chickpeas. The sweet juice of roasted lemon.

Serves 2

chickpeas 1 × 400g can
shallots, medium 6
olive oil 4 tablespoons
a lemon
thyme 6 bushy sprigs

butter beans 1 × 400g can
tahini 2 teaspoons
a small orange
radicchio or red chicory 400g

Set the oven at 200°C/Gas 6. Drain the chickpeas in a sieve. Peel the shallots, halve them lengthways and put them into a roasting dish, cut side down. Tip the chickpeas into the roasting tin, then trickle with the olive oil and grind over a little black pepper and salt.

Cut the lemon in half, squeeze lightly over the shallots, then tuck the lemon halves into the tin. Pick the thyme leaves from their stems, scatter them over the chickpeas and shallots, then roast for about thirty minutes, until the shallots are soft and their undersides are golden.

Lift the roasting tin onto the hob and transfer the shallots to a serving dish, leaving the chickpeas in the tin. Place the tin over a moderate heat and tip in the butter beans, drained of their canning liquor. Squeeze the roast lemons over the beans and chickpeas, then stir in the tahini and the juice of the orange.

Continue cooking until all is hot, then return the shallots to the roasting tin, tear the radicchio leaves into large pieces and fold together before transferring to a serving dish.

• It is useful to have a stash of roasted chickpeas around. You can make them and set them aside for a day or two, to be used as you wish: tossed as above with bitter leaves and citrus or folded into a salad of young spinach leaves or piled onto thick olive-oily toast.

FENNEL, CREAM, PINE KERNELS

Aniseed, citrus, crumbs and cream.

Serves 2

fennel 750g	parsley leaves 10g
olive oil	zest of 1 orange, finely grated
double cream 250ml	pine kernels 25g
fresh, white bread 45g	

Set the oven at 200°C/Gas 6. Trim the fennel, then slice into pieces no thicker than a pencil. The thicker you slice them, the longer they will take to cook. Warm a thin layer of olive oil in an ovenproof frying pan, then add the fennel, a few pieces at a time if there isn't room for them all at once, and let them cook until pale gold. Remove from the heat and pour in the double cream.

Tear up the white bread and place in the bowl of a food processor. Add the parsley leaves, the grated zest of the orange, a little salt and black pepper and the pine kernels. Process briefly to give a rough-textured crumb mixture.

Scatter the seasoned breadcrumb mixture over the surface of the cream. Trickle a little olive oil over the crumbs. Bake for twenty to twenty-five minutes until the cream is bubbling around the edges.

• Cooking fennel in this way softens the vegetable's aniseed notes. You can also make this recipe with celery instead of fennel, cutting the ribs into short lengths and baking with the cream as above.
• I sometimes bake small, floury potatoes at the same time, then spoon the fennel cream over the baked potato, mashing the aniseed-scented cream into the soft flesh of the potato.

LENTILS, SWEET POTATO, TOMATOES

A deep dish of comfort.

Serves 4

split yellow lentils 250g
ground turmeric 1 teaspoon
vegetable stock 750ml
sweet potatoes 750g
onion 1
olive oil 2 tablespoons
ginger 60g
garlic 2 cloves

cumin seeds ½ teaspoon
green cardamom pods 10
chilli flakes ½ teaspoon
curry leaves 12
ground cayenne 1½ teaspoons
chopped tomatoes 1 × 400g can
a little melted butter or oil,
    for baking

Wash the lentils in cold water, until the water is no longer milky. Tip them into a medium-sized saucepan with the turmeric and stock and bring to the boil. Lower the heat, partially cover with a lid and simmer for ten minutes, until most of the stock has been absorbed and the lentils are almost tender.

Peel the sweet potatoes and cut them into slices about a half a centimetre in thickness. Put them in a steamer basket or colander over a pan of boiling water, covered by a lid. Steam for about seven to eight minutes until soft to the point of a knife. Remove the pan from the heat.

Peel and roughly chop the onion. Warm the oil in a deep pan over a moderate heat, add the onion and cook until it is soft and pale gold – a matter of twenty minutes. Peel and grate the ginger. Peel and finely slice the garlic, then stir into the onion with the ginger and cumin seeds. Continue cooking for two or three minutes. Crack open the cardamom pods, extract the seeds, grind to a coarse powder, then stir into the onion together with the chilli flakes, curry leaves and ground cayenne. Add pepper and half a teaspoon of salt, stirring for a minute or two. Set the oven at 200°C/Gas 6. *(continued)*

Put the lentils over a moderate heat, stir in the tomatoes, then mix in the onion and remove from the heat.

Carefully lift the sweet potatoes from the steamer with a palette knife. Using a baking dish measuring approximately 24cm, place half the lentil mixture in the base, then add half the sweet potatoes over the surface, followed by a second layer of lentils and then the remaining sweet potatoes. Brush with butter or oil. Bake for thirty to forty minutes until bubbling around the edges and lightly browned on top.

• One of those dishes that is somehow even better the next day, when slowly reheated in a moderate oven.

MUSHROOMS, CHICKPEAS, TAHINI

A mushroom as thick as beefsteak. A silky purée.

Serves 2

large 'portobello' mushrooms 2
olive oil 8 tablespoons
garlic 2 cloves, peeled
ground sumac 2 teaspoons
juice of half a lemon

chickpeas 1 × 400g can
tahini 2 tablespoons
thyme leaves 1 tablespoon
black sesame seeds 1 tablespoon
white sesame seeds 1 tablespoon

Set the oven at 200°C/Gas 6. Cut out the stalks from the mushrooms, then place the mushrooms gill side up on a baking tray. Score the inside of each mushroom with the tip of a knife – it will allow the oil to penetrate – then pour one tablespoon of olive oil into each.

Use a pestle and mortar to crush the garlic, then pound in four tablespoons of the olive oil, the sumac, lemon juice and a little salt. Drain the chickpeas, then mash half into the oil and garlic paste. Stir in the tahini, thyme leaves and half of both the sesame seeds.

Fill the mushrooms with the chickpea paste then cover each with the reserved whole chickpeas. Finally, trickle over the last of the olive oil and scatter with the reserved sesame seeds. Bake for about thirty minutes.

• The best mushrooms for these are the very large portobello mushrooms with upturned edges to hold the filling.

ONIONS, TALEGGIO, CREAM

Caramelised onions. Melting cheese.

Serves 4

onions, medium 8	butter 60g
bay leaves 3	thyme 8 sprigs
black peppercorns 12	double cream 250ml
olive oil 3 tablespoons	Taleggio 200g

Bring a large, deep pan of water to the boil. Set the oven at 200°C/Gas 6. Peel the onions, keeping them whole. Add the bay leaves and peppercorns to the boiling water, then lower in the peeled onions. Leave the onions at a brisk simmer for about thirty minutes, till they are soft and yielding.

Lift the onions out with a draining spoon and put them into a baking dish. Pour the olive oil over the onions, add the butter, the leaves from half the thyme sprigs and a little black pepper, then bake for about thirty minutes, basting once or twice as they cook.

Warm the double cream in a small saucepan over a low heat. Cut the Taleggio into small pieces, then leave to melt in the cream, without stirring, adding the remaining sprigs of thyme.

Serve two onions per person, together with some of the Taleggio and thyme sauce.

• I sometimes serve this with a tangle of boiled noodles, or a spoonful of gnocchi to stir into the cheese sauce. The onions and their sauce are wonderful on thick slices of sourdough toast too.

PARSNIPS, SHALLOTS, GOAT'S CURD

Sweet, crisp and chewy parsnips. Thick and glossy gravy.

Serves 4

parsnips, medium 6	large (banana) shallots 4
olive oil 160ml	plain flour 2 tablespoons
rosemary 6 sprigs	thyme 10 sprigs
dried porcini 15g	goat's curd 100g

Set the oven at 200°C/Gas 6. Peel the parsnips, cut them in half lengthways, then in half again. Put them into a roasting tin with 100ml of the olive oil, the rosemary and a little salt and black pepper. Let the parsnips roast for an hour, turning them halfway through.

Bring a litre of water to the boil in a deep saucepan. Add the porcini, cover with a lid and continue cooking on a very low heat for twenty minutes.

Peel the shallots and cut them into thick slices, then cook them in the remaining oil for fifteen to twenty minutes over a moderate heat till soft. As they turn translucent and pale gold, add the flour and continue to cook, stirring regularly, for three minutes.

Pour the porcini and their broth into the shallots, a little at a time, stirring almost continuously. Remove the leaves from the thyme and stir in. Turn up the heat and let the sauce bubble for five minutes or so until you have a rich, quite thick gravy.

Remove the parsnips when they are golden and lightly crisp, spoon over the shallot and porcini gravy, and serve with the goat's curd.

• The shallots need to cook until they are golden and you can crush them between your finger and thumb, but stop well before they become nut brown and caramelised. Sticky, pale gold and translucent is perfect.

PARSNIPS, SMOKED GARLIC, FETA

Smoky, salty, sweet and toasted.

Serves 2

red onions 2	smoked garlic a whole head
swede 600g	olive oil 6 tablespoons
parsnips 2	fennel seeds 1 teaspoon
carrots, medium-sized 4	yellow mustard seeds 1 teaspoon
thyme 6 bushy sprigs	feta 200g

Set the oven at 200°C/Gas 6. Peel the red onions, cut them in half, then into thick segments. Peel and thickly slice the swede, then cut each slice into pieces roughly 4cm long. Do the same with the parsnips. Scrub the carrots, then cut into thick pieces and mix with the onions, swede and parsnips in a large roasting tin.

Tuck the thyme and smoked garlic amongst the vegetables, then pour over the olive oil. Roast for twenty-five minutes, then turn the vegetables over in the roasting tin and cook for a further twenty minutes. Remove the garlic from the roasting tin. Scatter the fennel and mustard seeds over the vegetables and return to the oven for a further twenty minutes till all is sizzling and the vegetables are knifepoint tender.

Squeeze the smoked garlic flesh from its papery skin, crush it to a paste with a pestle and mortar or using a spoon, then crumble the feta into it. Remove the roasted vegetables from the oven, then toss gently with the garlic and feta.

• A fine way to utilise the swede that turned up in the vegetable box. You could use pretty much any root vegetables in this, including young turnips or celeriac. If using the latter, make sure to baste regularly.

POTATOES, BRUSSELS SPROUTS

A cure for everything.

Serves 2

baking potatoes 4
garlic a whole head
double cream 250ml

Brussels sprouts 300g
olive oil 3 tablespoons

Prick the skins of the potatoes here and there with a fork. Bake them for about an hour at 200°C/Gas 6, until their skins are crisp and the inside is fluffy. Put the garlic in the oven and roast for forty-five minutes, till the cloves are soft inside.

Remove the garlic from the oven, squeeze the cloves from their skins and crush the flesh to a paste with a fork. Put it in a saucepan, pour in the cream and bring to the boil, then remove from the heat.

Quarter the Brussels sprouts. Warm the oil in a shallow pan, add the sprouts and let them brown lightly. Take the potatoes from the oven, slice off their tops and scoop the flesh into the pan with the sprouts. Fold the vegetables together, then stir in the warm garlic cream. Season with salt and black pepper, then spoon the filling back into the empty potato skins.

Return the potatoes to the oven for ten minutes, till piping hot.

• There isn't much I haven't stuffed into a baked potato in my time. From smoked mackerel and cream to crisp bacon and its fat, but a big fat tatty responds to other vegetables too. Try spooning over greens that you have sautéed in olive oil and lemon, a purée of green peas and mint, fried mushrooms with garlic and butter and mint, or cauliflower cooked with cream and Parmesan.

• The Brussels sprouts work well with a baked sweet potato too. They probably shouldn't, but the slight bitterness of the sprouts is surprisingly happy with the caramelised flesh of the sweet potato.

POTATOES, SWEET POTATOES, CREAM

The peace of potatoes and cream. The warm glow of spice.

Serves 4

butter 20g
onions, large 2
olive oil 5 tablespoons
baking potatoes 500g
sweet potatoes 500g

mild paprika 2 teaspoons
yellow mustard seeds 2 teaspoons
chilli flakes 2 teaspoons
double cream 750ml

Lightly butter a shallow, ovenproof 30cm dish. Set the oven at 160°C/Gas 3. Peel the onions and slice them into 1cm-thick rings.

Warm the olive oil in a deep pot, add the onions and let them cook until golden and soft. Expect them to take a good twenty-five to thirty minutes, with the occasional stir.

Peel and cut the baking potatoes into 0.5cm thick slices, then peel the sweet potatoes and cut them into slightly thicker slices, but no more than 1cm.

When the onions are golden, add the paprika, mustard seeds and chilli flakes, a good grinding of salt and black pepper, then pour the cream over and bring to the boil. Immediately remove from the heat.

Layer the potatoes of both varieties and the paprika cream in the buttered dish, then bake for an hour and half until bubbling and golden.

• Dauphinoise, updated with sweet potatoes and a generous hit of spice. Some salad leaves are needed with this, I think, something crisp, like chicory and iceberg lettuce or frisée with watercress. A sharp or piquant dressing would be my first choice – I'm thinking of red wine vinegar, peppercorns in brine, capers or something citrussy.

POTATOES, TAHINI, THYME

Roasties. Sesame. Lemon.

Serves 2

yellow fleshed potatoes 800g
olive oil 5 tablespoons
thyme 8 sprigs
garlic cloves 2
fennel seeds 2 teaspoons
spring onions 2

For the dressing:

juice of a lemon
tahini 1 tablespoon
olive oil 2 tablespoons

Set the oven at 200°C/Gas 6. Without peeling them, cut the potatoes into 'roast potato'-size pieces. Boil them in deep, salted water till just tender, about fifteen minutes.

Drain the potatoes, put them on a roasting tin or tray with the olive oil, thyme sprigs and unpeeled garlic and scatter with the fennel seeds. Turn the potatoes so they are coated, then roast for about fifty minutes until golden and crisp, removing the garlic after thirty.

Cut the spring onions into 3cm lengths and add to the roasting tin for the final ten minutes. Squeeze the garlic from its skin, then crush the soft, golden flesh with the lemon juice and tahini. Stir in the oil from the roasting tin. Should the dressing need it, add a little more olive oil to bring it to a smooth, pourable consistency.

Divide the potatoes between plates and spoon over the tahini dressing.

• I have yet to finish a jar of tahini. I usually need a couple of tablespoons for a special recipe then don't use any more for weeks or even months. Finding an everyday recipe that uses it is something of a gift and may help with getting through an entire jar before its sell-by date.

• This is one of my favourite recipes in the book, and I would be happy to eat it as a main dish, with salad leaves on the side.

POTATOES, TOMATOES, HORSERADISH

'Toast-rack' potatoes, spicy tomato sauce.

Serves 3

potatoes, large 6	garlic 6 cloves
olive oil 6 tablespoons	fresh horseradish, grated
tomatoes 800g	6 tablespoons

Set the oven at 200°C/Gas 6. Place the potatoes flat on a chopping board, then slice them at 0.5cm intervals, cutting almost through to the wood. It becomes immediately obvious why they are sometimes called toast-rack potatoes.

Place the potatoes in a roasting tin, trickle with some of the olive oil and roast for about forty-five minutes till crisp and golden. Meanwhile put the tomatoes and unpeeled garlic cloves in a small roasting tin with the remaining olive oil, grind over a little salt and black pepper, then place in the oven. Roast the tomatoes for about thirty-five to forty minutes, until they are soft and lightly browned.

Remove the tomatoes and garlic from the oven. Pop the garlic from its skin and add the flesh to the tomatoes. Crush the tomatoes and garlic with a fork, stirring in the fresh horseradish as you go. When the potatoes are crisp and golden, remove them from the oven and serve with spoonfuls of the crushed tomato.

• Some people find it easier to 'hasselback' the potatoes by placing each one in the hollow of a wooden spoon and slicing downwards, thus avoiding cutting through to the chopping board. It's a good trick.

PUMPKIN, CHICKPEAS, ROSEMARY

Roasted pumpkin. Smooth, silky mash.

Serves 4

pumpkin, skin on 1kg	*For the hummus:*
garlic 4 cloves	chickpeas 2 × 400g cans
rosemary 10 sprigs	juice of a small lemon
thyme 8 bushy sprigs	olive oil 150ml
a little olive oil	parsley leaves 10g
butter 75g	pink peppercorns 2 teaspoons

Set the oven at 200°C/Gas 6. Remove the seeds and fibres from the pumpkin, then cut the flesh into eight wedges. Lightly oil a baking tin (I like to cover mine with kitchen foil for easier cleaning) and lay the wedges down in a single layer. Tuck in the unpeeled garlic cloves. Season with salt, black pepper and the sprigs of herbs, then moisten with olive oil. Dot the butter in small lumps over the pumpkin and roast for forty-five minutes, until the squash is golden brown in colour and the texture is soft and fudgy.

Drain the chickpeas and bring them to the boil in deep water. Turn the heat down a little and let them simmer for eight to ten minutes till thoroughly hot. Squeeze the roast garlic cloves out of their skins. Drain the chickpeas again, then tip them into the bowl of a food processor. Add the lemon juice and garlic and process, pouring in enough of the oil to produce a soft, spreadable cream.

Chop the parsley and peppercorns together, then moisten with a tablespoon of olive oil. Spoon the hummus on to a serving dish, place the roasted pumpkin pieces on top, then scatter over the parsley and peppercorns and serve.

(continued)

• If you don't have pink peppercorns (they are hardly a kitchen essential), then I can recommend those dark green bottled peppercorns instead. They have both warmth and piquancy and are useful as a seasoning for anything containing beans or cheese. The brine they come in is handy, too. Just a few drops will perk up a salad dressing or a vegetable or bean purée.

PUMPKIN, COUSCOUS, DATE SYRUP

Sweet, fudgy flesh. Tart, bright dressing.

Serves 4

pumpkin 1kg (unpeeled weight)
olive oil 60ml
za'atar 1 tablespoon
dried chilli flakes 1 teaspoon
couscous 65g
coriander 20g
parsley leaves 10g

For the dressing:

garlic 1 clove
date syrup 4 teaspoons
olive oil 5 tablespoons
juice of half a lemon
grain mustard 1 teaspoon

Set the oven at 200°C/Gas 6. Peel the pumpkin, remove and discard the fibres and seeds, then cut the flesh into 2cm-thick slices. Place the slices on a baking sheet. Mix together the 60ml of olive oil, za'atar and chilli flakes, then spoon over the pumpkin. Bake for thirty-five to forty minutes till tender and translucent.

Bring a kettle of water to the boil. Put the couscous in a heatproof bowl, pour over enough of the hot water to cover and set aside.

Make the dressing: crush the garlic and mix with the date syrup, olive oil, lemon juice and grain mustard. Chop the coriander and the parsley.

When the couscous has soaked up the water, separate the grains with a fork, then fold in the parsley and coriander and season with salt and pepper. Remove the pumpkin from the oven, place the couscous on a serving plate with the pumpkin, then trickle over the dressing.

• Butternut squash is a handy year-round alternative to pumpkin and can be substituted here with ease. Butternut usually takes slightly less time than a variety such as Crown Prince, with its dense flesh. The trick as always with members of the squash family is to cook them until the golden interior is almost translucent and easily crushable with the back of a spoon.

PUMPKIN, MUSTARD, CREAM

Soft squash. The warmth of mustard.

Serves 3

pumpkin, skin on 2kg
olive oil 4 tablespoons
chicken or vegetable stock, hot,
   1 litre

double cream 200ml
grain mustard 1 tablespoon
smooth Dijon mustard
   1 tablespoon

Cut the pumpkin in half and each half into three large wedges. Remove and discard the seeds and fibres. Warm the olive oil in a large, deep-sided roasting tin over a moderate heat. Lightly brown both cut sides of each wedge of pumpkin, turning them over as each one colours using kitchen tongs.

Set the oven at 200°C/Gas 6. Lay the wedges of pumpkin on their side then pour over the hot chicken or vegetable stock and seal the roasting tin with kitchen foil. Bake for forty-five minutes, then remove the foil, turn the pieces of pumpkin over and baste them thoroughly with the stock. Return to the oven and continue cooking for a further forty-five minutes. The pumpkin should be translucent, each slice heavy with stock.

Carefully lift the pumpkin from the stock and set aside on a warm serving dish. Place the stock over a fierce heat and let it reduce to about 200ml. Pour in the cream, then stir in the two mustards, a little at a time, until you have a warmth you like. Season with salt and black pepper.

Spoon the mustard sauce over the slices of pumpkin and serve.

• You can cook any of the golden-fleshed autumn squashes in the style of fondant potatoes. That is, lightly browned in a little oil or butter, then baked in stock until they become saturated with the liquor, almost glowing. A texture so soft and giving as to be almost like sorbet. Such a recipe takes patience but little hands-on work. *(continued)*

• You need to baste the squash with stock once or perhaps twice during its sojourn in the oven. Cooked in this way, pieces of pumpkin make a juicy side dish, but I enjoy them as a course in their own right, with the sauce above, where the cooking juices are bolstered with cream and mustard.

SWEET POTATO, JALAPEÑOS, BEANS

Sweet heat.

Serves 2

a large sweet potato	butter 50g
radishes 4	black-eyed beans 1 × 400g can
a jalapeño	mozzarella 100g

Set the oven at 200°C/Gas 6, pierce the skin of the sweet potato with a fork a few times, and bake it for about forty-five minutes to an hour till the inside is soft and melting.

Slice the potato in half and scoop out the filling with a spoon, keeping the skin as intact as possible (it will be a little fragile), and mash with a fork. Cut the radishes into matchsticks. Thinly slice the jalapeño.

Warm the butter in a large frying pan, add the jalapeño and fry for two minutes, then tip in the drained beans. Stir in the mashed sweet potato, season and cook for three or four minutes. Tear the mozzarella into pieces and add to the sweet potato, then spoon back into the potato skins.

Scatter the matchstick radishes over the top and serve.

• You can use a floury maincrop potato instead should you wish, but the sweet potato seems to sit particularly happily with the beans and the coolness of the mozzarella.
• Use whatever beans you have to hand. Haricot, cannellini and butter beans will all fit the bill.
• The radishes add a pleasing crunch to each mouthful of soft potato and cheese. Use mooli if you prefer, or cool, crisp cucumber.

SWEET POTATOES, TOMATOES

Soft, spiced mash. Sharp, sweet, roast tomatoes.

Serves 3–4

sweet potatoes 850g
cherry tomatoes on the vine 500g
olive oil 8 tablespoons
onions, medium 2

garlic 2 medium cloves
ground turmeric 2 level teaspoons
yellow mustard seeds 3 teaspoons

Set the oven at 200°C/Gas 6. Peel the sweet potatoes, cut them in half lengthways, then into thick chunks, as you might for boiling. Place in a steamer basket over a pot of boiling water, then cover with a tight lid and let them steam to tenderness – a matter of twenty minutes or so. Test them every five minutes with a skewer.

Put the tomatoes in a shallow baking dish, tossing them with half the olive oil and grinding over just a little salt. Roast for about twenty minutes until their skins have blackened. While the sweet potatoes steam and the tomatoes roast, peel and thinly slice the onions, and cook them in the remaining olive oil in a shallow pan until they are soft and golden. Peel and thinly slice the garlic and stir it into the softening onions. When the onions and garlic are soft and golden, scoop them out into a small bowl.

Drain the juices from the tomatoes into the pan used for the onions and place over a moderate heat. Stir in the turmeric, letting it sizzle for a moment, then add the mustard seeds and a little black pepper. As soon as the mustard seeds start to pop, remove from the heat. Mash the sweet potatoes with a potato masher, then stir in the turmeric, mustard seeds and cooking juices.

Spoon the spiced sweet potato mash on to plates, then add the cooked onions and garlic and the roast tomatoes.   *(continued)*

• As the tomatoes roast, their juices will leak into the olive oil, forming a sweet-sharp base for the mustard seeds and turmeric. A handful of curry leaves, tossed in with the mustard seeds, wouldn't go amiss. Let the leaves warm for a minute or two in the hot oil with the mustard seeds, just long enough to lightly infuse the dressing with their earthy warmth.

ON A PLATE

My father's generation wouldn't have countenanced salad on a winter's day. But our eating moves on. Winter leaves, particularly the chicories, come into their own in cold weather; apples and pears are at their juiciest on a crisp autumn day; nuts and mushrooms are some of the finest ingredients the season has to offer.

On a golden autumn afternoon, I will make a salad of soft, freckled leaves of lettuce, with blackberries and blue cheese, and toss them in a dressing of cider vinegar and walnuts. As the nights draw in, mushrooms will be marinated with orange juice and sherry vinegar, then sprinkled with toasted hazelnuts and toasted white breadcrumbs. In deep midwinter, beetroots will be baked and eaten warm with white-tipped radishes, mint and segments of blood orange.

I will not forgo my love of leaves, tufts of watercress and ice-crisp raw vegetables just because there's a fire in the hearth. A plate of salad leaves is not just for summer. Whilst they are unlikely to form the main course, arrangements of raw vegetables sit comfortably aside a bowl of deep ochre soup and a lump of open-textured bread torn from a sourdough loaf.

The trick is in the dressing. Walnut and sesame oils, pickling liquors and cider vinegars, toasted nuts, grain and smooth mustards all marry with the ingredients of the season – the russet-skinned apples, fat juicy pears, hot mustardy leaves and crunchy white celery.

The fragile salad leaves of summer give way to shreds of red and white cabbage, onions fast-pickled in white wine vinegar, tight bulbs of chicory, thin slices of fennel (marinate them in lemon juice to soften their aniseed notes) and finely sliced raw kale. There is a robustness to a composition of winter vegetables that is very different to the immature and tender leaves of summer. Stems are thicker and more crunchy, leaves have more bite, flavours are more assertive.

Serving food on a plate does give the ingredients a chance to shine, but I think we should avoid the temptation to tinker just to make something look attractive. By all means hand over a tempting-looking plate of food, but nothing good will come from plated arrangements. Letting food fall naturally into place will always look more appetising than food that has been 'arranged'.

APPLES, BLUE CHEESE, WALNUTS

Crisp apples and toasted nuts. The colours of autumn.

Serves 2
For the dressing:
blackberries 65g
cider vinegar 4 tablespoons
walnut oil 3 tablespoons
olive oil 3 tablespoons

walnut halves 75g
a large apple
blackberries 60g
a winter lettuce such as Castelfranco
blue cheese such as Shropshire Blue,
    Stilton or Stichelton 250g

Make the dressing: put the blackberries, cider vinegar and the oils into a food processor and blend for a few seconds, then tip into a salad bowl.

Toast the walnut halves in a dry pan till warm and fragrant, then add to the dressing. Quarter, core and slice the apple and gently toss with the walnuts, blackberries and dressing. Tear the lettuce into large pieces, teasing the heart leaves apart, then divide between two plates. Spoon the dressing over the leaves then crumble the cheese on top.

• Instead of the lettuce, use finely shredded cabbage or sauerkraut. Blueberries can replace the blackberries and pear can be used instead of the apple.

BEETROOT, BLOOD ORANGE, WATERCRESS

Bright flavours for a cold day.

Serves 2

beetroot, small raw 400g
olive oil 1 tablespoon
radishes 8
blood oranges 2
watercress 100g
mint leaves 10

parsley leaves a handful
pumpkin seeds 2 tablespoons

For the dressing:
blood orange juice 2 tablespoons
sherry vinegar 2 tablespoons

Set the oven at 200°C/Gas 6. Place a large piece of cooking foil in a roasting tin. Wash and trim the beetroot, taking care not to break their skins, and place them in the foil. Add the olive oil, 2 tablespoons of water and a grinding of salt and pepper, then scrunch the edges of the foil together to seal. Bake the beetroots for forty-five to sixty minutes till tender to the point of a knife, then remove and set aside.

Slice the radishes in half and put them in a mixing bowl. Slice the peel from the oranges, taking care to keep any escaping juice, remove the skin from the segments, then add to the radishes. Peel the beetroot and cut each one in half, then into quarters, and add to the mixing bowl.

Wash and trim the watercress. (I like to dunk it in a bowl of iced water for twenty minutes to crisp up.) Put the mint and parsley leaves into the orange and beetroot together with the trimmed watercress and pumpkin seeds.

Make the dressing: put the blood orange juice in a small bowl, add the sherry vinegar, season with salt and pepper then pour over the salad and toss gently.

MUSHROOMS, ORANGE, BREADCRUMBS

Bosky flavours. Toasted crumbs.

Serves 4

juice of a blood orange
juice of a lemon
olive oil 4 tablespoons
sherry vinegar 2 tablespoons
chestnut mushrooms 250g
black figs 8

For the crumbs:

hazelnuts, skinned 75g
white bread 50g
olive oil 4 tablespoons
chopped parsley a handful
oakleaf lettuce 1, small
green olives, pitted 4 tablespoons

Mix together the orange and lemon juices, olive oil and vinegar, then grind in a little salt and pepper. Thinly slice the mushrooms, put them into the marinade, then cover the bowl and set aside.

Roughly chop the hazelnuts, then toast them in a dry, shallow pan. Set aside in a bowl. Process the bread to coarse crumbs. Warm the olive oil in the same pan over a moderate heat, then add the breadcrumbs and toss them gently in the hot oil till pale gold. Stir in the chopped parsley and toasted hazelnuts and set aside.

Tear the lettuce into large pieces and divide between four plates or bowls, then add the marinated mushrooms, the toasted crumbs and lastly the figs, cut into quarters, and the olives, halved if you like.

• Add other seeds and nuts to the crumbs as you wish. When the crumbs are frying, add pumpkin or sunflower seeds, a sprinkling of chia, a handful of golden sultanas or some flaked almonds.
• Instead of the crumbs, use cooked couscous, quinoa or puffed rice.

RED CABBAGE, CARROTS, SMOKED ALMONDS

Crisp, crunchy, sour and smoky. A cabbage salad for a winter's day.

Serves 4

a red onion
malt vinegar 50ml
cider vinegar 75ml
yellow mustard seeds 1 teaspoon
red cabbage 450g
carrots 250g
a pear
smoked almonds a handful or two

For the dressing:

soured cream 150ml
pickling liquor from the onion
   4 tablespoons
Dijon mustard 1 teaspoon
grain mustard 1 teaspoon
poppy seeds 1 tablespoon

Peel the onion and finely slice into rings. Warm the vinegars, 120ml of water and the mustard seeds in a small saucepan, add half a teaspoon of salt and the onion. Bring to the boil, then remove from the heat, cover and leave for twenty-five minutes.

Finely shred the red cabbage. Peel the carrots, then slice them into long shavings with a vegetable peeler. Halve, core and slice the pear. Toss the cabbage, carrots and pear with a little of the onion pickling liquid.

Make the dressing: lightly beat the soured cream, onion pickling liquor, mustards and poppy seeds. Fold the dressing into the shredded cabbage, carrots and pear. Finally, add the smoked almonds and the pickled onions, drained of their remaining pickling liquor.

• Should smoked almonds prove elusive, make your own by mixing smoked salt with a little paprika and groundnut oil, then toast in a dry pan, before adding the whole almonds.

WITH A CRUST

You push your spoon through the crust. The pastry, potato, toasted cheese or crumbs give under its weight. A puff of aromatic steam. Steam scented with onions and bay, thyme or rosemary, tarragon or garlic. And then your spoon sinks into the depths below. There may be melting cheese or a cloud of sweet potato, there might be mushrooms, sprouts or silky shallots. Whatever, it will be something to nourish and soothe, warm the heart and calm the soul.

Of course, your crust may be below the filling. A layer of pastry supporting shallots and apple, or a fan of late autumn plums. Maybe some slices of banana glistening with maple syrup. The pastry may be puff or rich buttery shortcrust, it might have chopped thyme rubbed into it, or grated Parmesan.

Food cooked with a crust is both soft and crisp. I am not sure food can get better than this on a winter's night. A puff-pastry parcel filled with mashed sweet potato; a deep dish of filo pastry filled with almost-liquid melted Taleggio and toothsome greens. Layers of swede, mushrooms and Gruyère with a crumbly thyme-freckled wrapper of shortcrust; a herb crumble baked on top of a cream sauce in which sleep tomatoes, leeks and basil.

I delight in crusts that split here and there. Holes and tears and gaps where the filling peeps deliciously through. Sometimes it floods the pastry: something I always hope for but which happens all too rarely – a tantalising glimpse of what lies beneath. Be it creamy or simmering in its own juices, spiced or herbed, piping hot or served relaxed and at room temperature, the filling is all the better for having a little of the crust come with it.

I rarely put pastry both under and on top of the filling. The base often fails to crisp at all. And yet a single sheet of ready-rolled puff pastry can be draped over any vegetable stew and baked; a few sheets of filo, each brushed with melted butter (and, if you wish, crumbs or chopped thyme) can provide an instant tart base. For everyday eating I would go for a top or bottom crust, not both.

And although this chapter is complete in itself, it would be remiss of me not to point you in the direction of the pudding chapter, and the quickest banana and maple syrup tarts.

FILO PASTRY, CHEESE, GREENS

A tart to eat straight from the oven, while the cheese is still soft and melting.

Serves 4

purple sprouting broccoli 250g

sprout tops 150g

butter 90g

filo pastry sheets 7 (270g)

Taleggio 500g

Parmesan 150g

You will also need a metal baking dish or tart tin about 30cm in diameter.

Bring a large, deep pan of water to the boil. Set the oven at 200°C/Gas 6. Place a baking sheet or pizza stone in the oven. Slice each stem of purple sprouting broccoli in half lengthways. Finely shred the sprout tops. Lower the broccoli into the boiling water, leave for three or four minutes till the colour is bright, then lift out with a draining spoon and refresh in a bowl of iced water. Do the same with the shredded sprout tops.

Melt the butter in a small pan. Brush the base of the baking dish or tart tin with some of the butter, lay a sheet of pastry over the base and brush it with more butter. Repeat with the remaining sheets of pastry, letting them overhang the edges of the tin.

Tear the Taleggio into small pieces and grate the Parmesan. Drain the vegetables and dry on kitchen paper. Fill the tart case with the drained purple sprouting broccoli and sprout tops, then tuck the Taleggio amongst them and scatter over the grated Parmesan. Fold the overhanging pastry back over the edges of the tart, place on the heated stone or baking sheet and bake for about twenty to twenty-five minutes, till the pastry is crisp.

• Putting a metal baking sheet or pizza stone in the oven before you start is a sound idea. As the oven warms, the sheet will heat up, providing an extra hot base on which to bake your tart. Doing this will ensure a crisper base than if you simply place the tart on the oven shelf.

LEEKS, PARSNIPS, PASTRY

Bosky vegetables. Cream sauce. Crisp crust.

Serves 4

leeks, medium 2	smoked garlic 4 cloves
butter 45g	thyme 12 sprigs
chestnut mushrooms 200g	double cream 350ml
olive oil 4 tablespoons	puff pastry 325g
parsnips 400g	a small egg, beaten

Set the oven at 200°C/Gas 6. Cut the leeks into 2cm-thick rings and wash them thoroughly. Put the leeks in a saucepan with the butter, place a piece of greaseproof paper or baking parchment over the surface of the leeks, then cover tightly with a lid. Let the leeks steam in the butter over a moderate heat until tender, but not coloured.

Thickly slice the mushrooms. In a shallow ovenproof pan about 24cm in diameter, warm the olive oil, add the mushrooms and fry till golden. Meanwhile, peel the parsnips and cut into 1cm dice. Peel and thinly slice the garlic. Remove the mushrooms as they are ready, then add the parsnips to the pan, together with a little more oil if necessary. Let them brown lightly, adding the garlic after five minutes, then remove them and return the mushrooms to the pan.

When the leeks are tender, remove the lid and paper and add the leeks and parsnips to the mushrooms. Season the cream with salt, pepper and the sprigs of thyme, then pour over the vegetables, remove from the heat and leave to cool a little.

Roll the pastry out into a disc large enough to cover the pie. Lower the pastry into place over the vegetables, pressing it tightly over the edges. Brush with the beaten egg, then cut a small hole in the centre to let the steam out. Bake for twenty-five minutes till pale gold and the cream is bubbling.

LEEKS, TOMATO, PECORINO

Hearty but not heavy.

Serves 4
leeks 850g
butter 40g
cherry tomatoes 300g
basil leaves 10g
double cream 250ml

For the crumble:
plain flour 250g
butter 125g
Pecorino, grated 75g
thyme 10g
parsley 15g

Cut the leeks into 1cm-thick rounds, then wash very thoroughly in cold water. Put the leeks into a deep ovenproof casserole or saucepan with the butter, cover with a lid and let them soften over a moderate heat. A regular stir will stop them colouring. Cut the tomatoes in half and stir into the softening leeks, together with some sea salt and black pepper, letting them cook for about ten minutes until soft and juicy. Stir in the basil leaves, then the cream. When the cream has warmed, remove from the heat. Set the oven at 200°C/Gas 6.

Make the crumble: put the flour in a large mixing bowl, add the butter in small pieces and rub in with your fingertips. Stir in the grated cheese. Pull the leaves from the thyme and parsley, finely chop and stir in to the crumble. You can do this in seconds in a food processor. Sprinkle several drops of water over the mixture and shake the bowl back and forth until you have a mixture of large and small crumbs. Scatter the crumble over the tomato and leeks then bake for forty minutes till bubbling round the edges.

• Parmesan or a firm, extra mature Cheddar, grated finely, will do the job of the Pecorino if you prefer. Rosemary, finely chopped, can work in place of thyme, or you could use the dried herb mixture 'herbes de Provence'.

SHALLOTS, APPLES, PARMESAN

A classic tart. A savoury twist.

Serves 6

For the pastry:

plain flour 225g

butter 150g

egg yolk 1

thyme leaves 2 teaspoons

Parmesan, finely grated 4 tablespoons

banana shallots, medium 4

olive oil 2 tablespoons

apples 2

You will need a tarte Tatin tin or a metal-handled frying pan measuring 24cm in diameter.

Put the flour into the bowl of a food processor, add 120g butter in small pieces and process to the texture of fine breadcrumbs. Add the egg yolk, thyme leaves and 3 tablespoons of the grated Parmesan, process briefly, then transfer to a lightly floured board and bring together into a ball. Of course, you can do this by hand if you wish, rubbing the flour and butter together with your thumb and fingertips (a peaceful flour-dusted thing to do), then fold in the yolk, thyme and cheese. Wrap in baking parchment or place in a bowl and cover, and leave to rest in the fridge.

Set the oven at 200°C/Gas 6. Peel the shallots, then halve lengthways. Melt the remaining 30g of butter with the oil in the tin or frying pan over a moderate heat, then add the shallots, cut side down. Let them brown lightly, then turn to let the other side colour. Meanwhile, halve, core and slice each apple into eight segments. Remove the shallots from the pan, then add the apples, letting them soften and turn lightly gold. Scatter over the remaining tablespoon of grated Parmesan, then return the shallots.

Roll the pastry out to a good 3cm larger than the tin or frying pan. Lay the pastry over the shallots and apples, tucking in the overhanging dough. Bake for twenty-five minutes, until the pastry is pale biscuit-coloured and the butter is bubbling round the edges. Remove from the oven, leave to settle for ten minutes, then turn out onto a serving plate.

SWEDE, MUSHROOM, GRUYÈRE, THYME

Herb scented, rib-sticking filling. Crumbly pastry.

Serves 6

For the pastry:
plain flour 200g
rye or wholemeal flour 100g
thyme leaves 1 tablespoon
butter 150g
salt a pinch
an egg, beaten

For the filling:
a medium onion
butter 90g
swede 450g
chestnut mushrooms 250g
Gruyère 250g

Make the pastry: put both the flours and the thyme leaves in a large bowl, cut the 150g of butter into small pieces, then add it with the salt to the flour and rub together with your fingertips until you have a breadcrumb-like consistency. Introduce enough cold water to produce a soft, rollable dough. Alternatively, you could make this in a food processor. Turn out onto a floured work surface, pat into a ball, flatten the top, then wrap in baking parchment and refrigerate for thirty minutes.

Set the oven at 200°C/Gas 6. Place an upturned baking sheet or pizza stone in the oven to heat up. Peel and roughly chop the onion then let it soften in 30g of the butter over medium heat in a large, wide pan. Peel the swede, cut it in half lengthways, then cut each half into quarters. Slice each quarter thinly. Remove the onion from the pan – it should be pale gold and translucent. Add the pieces of swede to the pan, adding more butter if necessary, turning them as they brighten and soften. Thinly slice the mushrooms, then add them to the pan, with the remaining butter, and cook till pale gold. Grate the Gruyère into a large bowl, then add the swede, onions and mushrooms. Season generously with black pepper.

On the floured board, roll the pastry out into a large, rough-edged circle approximately 30cm in diameter and transfer to a parchment-lined

(continued)

or flour-dusted baking sheet. Pile the filling in the centre of the pastry, leaving a wide gap of bare pastry round the edge. Fold the edges of the pastry over the filling, leaving the middle open. Brush the edges with beaten egg, then place the whole baking sheet on top of the hot sheet or stone in the oven and cook for twenty-five minutes, till the pastry is pale gold and the filling has crisped on top. Leave to settle for ten minutes before slicing.

• This is rather good cold too, and can be sliced and tucked inside a lunchbox or taken on a picnic.

SWEET POTATO, PUFF PASTRY

Crisp pastry. Soft, mildly spicy stuffing.

Serves 4

sweet potatoes 850g
harissa paste 2 teaspoons
ras el hanout 1 teaspoon
puff pastry 325g

an egg
black sesame seeds a couple
of pinches

Set the oven at 200°C/Gas 6. Place an empty baking sheet in the oven to warm. (It will help the base of your pastry crisp.) Put a pan of water on to boil and balance a steamer basket or colander over it. Peel the sweet potatoes, cut them into large pieces and steam for ten to fifteen minutes until tender. Remove from the steamer, then mash with a potato masher or food mixer till smooth. Season with sea salt, the harissa paste and the ras el hanout and set aside to cool.

Roll the pastry out to a rectangle measuring 35 × 23cm, then transfer to a baking sheet, turning the pastry so the long side is facing you. Spoon the cooled sweet potato on the right-hand half, leaving a 2cm rim around the edge uncovered, then smooth the surface flat.

Break the egg into a small bowl and beat lightly with a fork to mix yolk and white. Brush the bare edge generously with the beaten egg, then fold the left-hand side over the right, as if you were closing a book, and press firmly around the edges to seal.

Score a trellis pattern on the top of the pastry, then brush with more of the beaten egg. Scatter the black sesame seeds over the top, then bake for forty minutes on the hot baking sheet till golden brown.

• The filling should be soft but not wet. Should you choose to boil the sweet potato instead of steaming, make sure to drain thoroughly, and leave uncovered for ten minutes to allow the steam to escape. *(continued)*

• I have also made this with maincrop, floury-fleshed potatoes, adding a handful of grated cheese after mashing. A deeply satisfying pie for a cold day.

With a crust 161

WITH A LADLE

A ladle is the spirit of generosity. Deeply comforting, capacious, motherly. A ladle is a spoon that is expecting friends for dinner. (Even the largest kitchen spoon feels mean beside the smallest ladle.) Generosity that goes beyond mere quantity. You dip your ladle deep into a pot of soup or stew and lift out enough to fill the hungriest of diners, a bowl of steaming savour and sustenance. Full to the brim, overflowing with welcome.

And yet what we serve from a ladle is often humble – a soup of beans and woody herbs; a piping hot onion broth; a scented stew of noodles and spices. No matter how basic the ingredients, the food we share from this piece of bent metal is at once hearty and wholesome.

A ladle doesn't do dainty, twee or tight-fisted. It doesn't do lonely. You don't often serve your own dinner with a ladle. Like a pot of tea, this is a piece of equipment that is about sharing something with others. You may ask everyone to help themselves to a spoonful of pie. To cut themselves a slice of bread. But it is invariably the cook who wields the ladle.

Some foods ask to be served from this bowl-on-a-handle. Broth obviously, smooth vegetable purées too, but also lumpy soups and wannabe stews. But nothing with a crust, which needs a wide spoon to ensure the pastry or crumb topping stays on top. A ladle is there for a soup of black-eyed beans and greens; small seed-shaped pasta with grated cheese and herbs; a creamy purée made from a cauliflower or pumpkin and, of course, stock.

My first ladle, taken from a restaurant that had more than it needed, was too wide. The soup gushed over the sides of the narrow-mouthed bowls I had at the time. The second, a gift from a friend (which surely belonged with a silver-plate tureen in a grand restaurant), had a flamboyant kink in the handle that made it uncomfortable to hold. It twisted as you poured. The next had a handle too upright from the bowl and made aiming the contents difficult. It left a few stains behind.

The handle should sit at a slight angle to the bowl, so it pours comfortably as you turn your hand to the left (assuming you are right-handed), which discounts most of those sold in catering supply shops, whose handles seem to sit straight against the bowl. And they must have a hook or a sharp bend at the end, so you can hang them by the cooker, which is where they are going to be of most use. I don't think it should be stainless steel, too clinical and cold-hearted. The one I have settled with is made of white enamel, a little chipped now, and has been with me for so long it feels like an old friend.

The right ladle, like the right knife, is the one that feels made for your hand. Ideally, it should look as much at home at the table as it does in the kitchen. A cheap one will probably work as well as any. Try before you buy. Hold it aloft in the shop and scoop bowlfuls of air into an imaginary soup bowl. How does it feel? A good ladle is a friend for life, and will be a friend to your friends, who are sitting patiently at the table, waiting for you to turn up with it in your hand.

BEETROOT, LENTILS, GARAM MASALA

Soft, soothing spice.

Serves 4

a medium onion
olive oil 3 tablespoons
garlic 2 cloves
chilli flakes 2 teaspoons
yellow mustard seeds 1 teaspoon
vegetable stock 800ml

split orange lentils 300g
cooked beetroot 250g
butter 50g
garam masala 4 teaspoons
natural yoghurt 4 tablespoons

Peel the onion and slice into rings. Pour the oil into a shallow pan, add the onions, then fry to a deep, golden brown – a good twenty-five minutes – stirring from time to time. Peel and thinly slice the garlic, adding it to the onions halfway through cooking. Stir in the chilli flakes and mustard seeds and continue cooking for a couple of minutes.

Bring the stock to the boil in a medium-sized saucepan, tip in the lentils, lower the heat slightly, then let them cook for fifteen minutes till soft.

Roughly chop the beetroot and fold into the lentils, adding the butter and garam masala, then check the seasoning, adding salt and black pepper as necessary. Continue cooking for three or four minutes, then spoon into bowls.

Spoon the onions and yoghurt on top, stirring each bowl thoroughly to mix the onions, mustard and chilli into the dhal as you eat.

• Mild and slightly sweet, this lentil dish becomes interesting when you stir in the yoghurt and spiced onions. Increase the chilli level to suit your taste, or introduce a little mashed fresh ginger.

BEETROOT, SAUERKRAUT, DILL

The sweetness of beets. The sourness of cabbage. The heat of horseradish.

Serves 2

vegetable stock 600ml
cooked beetroot 500g
sauerkraut 100g
dill, chopped 4 tablespoons

soured cream 150ml
fresh horseradish a thumb-
  sized piece

Bring the stock to the boil in a large saucepan. Put the beetroot in the bowl of a food processor and process to a coarse purée. Stir the beetroot into the stock and season generously with black pepper and a little salt.

Mix the sauerkraut with the dill. Ladle the beetroot and stock into deep bowls. Divide the sauerkraut between them, then spoon over the soured cream. Finely grate about a teaspoon of horseradish (to taste) over the surface of the soup.

• I use whichever stock is to hand for soups such as this. Home-made if I have it, otherwise bought readymade or made up from bouillon powder. They all have their merits.
• Sauerkraut is one of those ever-present ingredients in my fridge, like yoghurt, lemons and tsukemono, the jewel-hued Japanese pickles. I buy it in jars or packets and dip into it on an almost daily basis. It makes a refreshing accompaniment to almost anything you care to mention.

BLACK-EYED BEANS, ROSEMARY, KALE

Therapy for a cold, wet winter's night.

Serves 4

a large onion
a large, fat carrot
rosemary 2 sprigs
celery 2 sticks
garlic 2 cloves
olive oil 3 tablespoons
a large tomato
black-eyed beans 2 × 400g cans

vegetable stock 750ml
bay leaves 2

To finish:
kale or Brussels sprout tops 150g
parsley leaves a good handful
olive oil

Peel and roughly chop the onion and the carrot. Pull the needles from the rosemary and chop finely. Remove the leaves from the celery and reserve, then chop the stalks into small pieces. Peel and finely chop the garlic.

Warm the olive oil in a wide, deep saucepan. When the oil is hot, add the onion, carrot, rosemary, celery and garlic and cook over a low to moderate heat for about fifteen minutes, till the vegetables are starting to soften. Roughly chop and stir in the tomato.

Tip the beans in with the vegetables, rinsing them first if you wish, then pour in the vegetable stock (water at a push) and bring to the boil. Add the bay leaves, lower the heat, then simmer for fifteen minutes until thoroughly hot. Season generously with salt and black pepper. Ladle one-third of the mixture into a blender, process till smooth, then stir back into the soup.

Pick over the kale or sprout tops and tear into pieces that will fit comfortably in the bowl of a spoon. Dunk the kale or sprout leaves under the surface of the soup and leave to soften for two or three minutes.

Roughly chop the parsley and the reserved celery leaves. Ladle the soup into bowls, scatter the surface with the chopped celery and parsley leaves and finish with a trickle of olive oil.

BRUSSELS TOPS, BLUE CHEESE

The goodness of greens. The savour of cheese.

Serves 4

milk 600ml
Stilton 200g
onions 2
butter 30g
sprout tops 350g

To finish:

the reserved sprouts (from the tops)
butter 30g
Stilton 100g, crumbled

Bring the milk to the boil, then remove from the heat. Crumble the Stilton into the milk, cover with a lid and leave to infuse. Peel the onions, then roughly chop them.

Melt the butter in a large, deep pan and add the chopped onion. Fry gently over a moderate heat for about ten minutes, until the onion is glossy and sweet smelling, stirring regularly. Pull the sprout leaves from the main stem, setting aside any whole tiny sprouts you may find. Remove the stems from the leaves and chop them, then stir into the onions.

Place the leaves on top of one another, roll them up, then slice into thin shreds and stir into the onion. Continue cooking for five minutes until the leaves are bright green and wilted. Pour in the milk and cheese mixture and bring almost to the boil. Ladle into a blender, taking care not to overfill (you will need to do this in two batches) and process to a thick but far from totally smooth soup. Check the flavour, adding black pepper if necessary. The soup is unlikely to need salt.

To finish, cut the reserved sprouts in half and fry them for a minute or so in the butter until they are bright and crunchy. Toss them with the crumbled Stilton and divide between the bowls.

CELERIAC, HORSERADISH, PUMPERNICKEL

A soup to soothe the soul.

Serves 4

celeriac 1kg

vegetable stock 400ml

milk 600ml

chives 25g

olive oil 125ml

grated horseradish root 15g

pumpernickel or other rye
 bread 100g

Peel and roughly chop the celeriac, then place in a deep pan with the stock and milk and bring to the boil. Lower the heat and simmer for twenty minutes.

Whilst the soup is simmering, chop the chives and put them in a blender with the olive oil. Blitz to a bright green dressing.

Purée the celeriac mixture, in two or three batches, using a blender. If you have a stick blender, then process it directly in the pan. Stir the horseradish into the soup and check the seasoning. Crumble the pumpernickel and toast briefly till lightly crisp in a dry, shallow pan.

Ladle the soup into bowls, pour in the chive oil and scatter over the pumpernickel.

• Celeriac is one of the more undersung vegetables, but has plenty going for it. I like it shredded and stirred into a mustardy mayonnaise, but also coarsely grated and fried in olive oil and butter until golden and sizzling and served with a spritz of lemon.

CHEDDAR, CIDER, MUSTARD

Deep bowls of velvety soup.

Serves 6

onions, medium 2	plain flour 45g
butter 30g	vegetable stock 400ml
olive oil a little	cider 350ml
carrots 2, medium	mustard 1 teaspoon
celery 1 stick	Cheddar 400g
milk 400ml	chopped parsley a small handful

Peel and roughly chop the onions. Warm the butter and oil in a deep saucepan over a moderate heat, then add the onions and cook till soft. Scrub and finely dice the carrots, finely dice the celery, then add to the softening onion and continue cooking for ten minutes or so till tender.

Warm the milk in a small saucepan and set aside. Stir the flour into the vegetables and continue cooking for two or three minutes, then add the milk and stir to a thick sauce. Pour in the vegetable stock and cider, bringing it to the boil, then lowering the heat and letting the mixture simmer for a few minutes. Stir in the mustard and check the seasoning. (It may want pepper, but probably very little salt.)

Grate the cheese and stir into the soup, leaving it to simmer (it is crucial it doesn't boil) for five minutes, until the soup has thickened. Add a handful of parsley as you bring it to the table and serve with bread.

• I suggest a mature, full-flavoured Cheddar such as Keens or Montgomery here. The stock should, if possible, be home-made, chicken or vegetable. Bread is pretty much non-negotiable here, either torn and placed in the base of the soup bowl, the soup ladled over it, or toasted and dipped into the soup's creamy depths in fat, jagged chunks.

FENNEL, CUCUMBER, MINT

A chilled soup for a warm autumn day.

Serves 4

fennel 400g
butter 75g
juice of half a lemon
cucumber 300g
radishes 12

pickle juice (from a jar of cornichons) or white wine vinegar 2 tablespoons
mint leaves 12
yoghurt 250ml
ice cubes

Thinly slice the fennel. Melt the butter in a deep pan, add 125ml of water, the lemon juice and fennel and cover with a piece of greaseproof paper or baking parchment. Cover with a lid and cook for about twenty minutes over a low to moderate heat, so the fennel steams rather than fries, cooking without colour. When fully tender, process to a purée in a blender.

Cut the cucumber in half lengthways, scrape out the seeds with a teaspoon, then grate the flesh coarsely. Grate the radishes and mix with the cucumber. Stir in the pickle juice or vinegar, and the fennel purée, then shred the mint leaves and stir them in, together with the yoghurt.

Ladle into bowls, then add a couple of ice cubes to each and, if you wish, a couple of radishes.

• A cool, bright-tasting soup given piquancy with a little juice from the pickle jar. I have also made this slightly thicker, using a strained yoghurt, and eaten it with dark rye bread.
• Don't be tempted to serve this soup hot – it curdles when heated. It is my belief that chilled soups must be thoroughly cold. With that in mind I store the soup in the fridge, then serve it with ice cubes dropped in. A brilliant answer as to what to eat on a bright, sunny autumn lunchtime.

MUSHROOMS, BUTTERNUT, SOURED CREAM

A bowlful of autumn.

Serves 4

butternut squash 1kg
shallots 6, medium
olive oil 5 tablespoons
sweet paprika 1 teaspoon
hot smoked paprika 1 teaspoon
vegetable stock 1 litre
soured cream 150g

For the mushrooms:

chestnut mushrooms 200g
olive oil 3 tablespoons
ginger 1 × 10g piece

Set the oven at 200°C/Gas 6. Peel the butternut, halve it lengthways and discard the seeds and fibres. Cut the flesh into large chunks, then place in a single layer on a baking sheet or roasting tin. Peel the shallots, halve them, then tuck them among the squash.

Mix the olive oil and the paprikas, then spoon over the squash, slide the dish into the oven and bake for about an hour, or until the squash is patchily browned but thoroughly soft and tender.

Warm the vegetable stock in a large saucepan. Add the squash and shallots to the stock, place over a moderate heat (use a few spoonfuls of the stock to deglaze the roasting tin if there are any interesting bits of roasted squash and shallot left), then season with salt and a little pepper and bring to the boil. Partially cover the pan with a lid and simmer for twenty minutes, until the squash is falling to pieces. Crush a few pieces of the squash into the liquid with a fork to thicken it slightly.

Slice the mushrooms thinly, then fry them in the olive oil in a shallow pan until golden. Peel the ginger and shred it into very fine matchsticks, then add it to the mushrooms and continue cooking for a minute or two. When all is golden and sizzling, remove from the heat.

Ladle the soup into deep bowls, then spoon over the soured cream and the mushrooms and ginger.

NOODLES, LENTILS, SOURED CREAM

A frugal, rich and sustaining soup-stew. My version of the Iranian *ash reshteh*.

Serves 4–6

onions 4	butter 40g
olive oil 3 tablespoons	linguine or Iranian *reshteh*
garlic 3 cloves	noodles 100g
ground turmeric 2 teaspoons	spinach 200g
chickpeas 1 × 400g can	parsley 30g
haricot beans 1 × 400g can	coriander 20g
small brown lentils 100g	mint 15g
vegetable stock 1 litre	soured cream 250ml

Peel the onions. Roughly chop two of them and thinly slice the others. Warm the olive oil in a large pan set over a moderate heat, add the two chopped onions and fry them for ten to fifteen minutes till soft and pale gold. Peel and thinly slice the garlic, then stir in with the ground turmeric and continue cooking for a couple of minutes.

Drain the chickpeas and haricots and stir into the onions together with the lentils and stock. Bring to the boil, then lower the heat and leave to simmer for 30 minutes, stirring occasionally.

Melt the butter in a shallow pan, then add the reserved sliced onions and let them cook slowly, with the occasional stir, until they are a rich toffee brown. This will take a good half an hour, maybe longer.

Add the noodles to the simmering beans. Wash the spinach, put it in a pan set over a medium heat, cover with a lid and leave it for three or four minutes until it has wilted. Turn occasionally with kitchen tongs. Remove the spinach and put it in a colander under cold running water until cool. Wring the moisture from the spinach with your hands, then stir into the simmering stew. Roughly chop the parsley, coriander and mint leaves and stir most of them into the onions and beans. *(continued)*

Fold in the soured cream, then ladle into bowls and fold in the remaining herbs and the fried onions.

• One of the dishes we ate time and again when in Iran, this is a bowl of deep, wholesome goodness. The soup is soothing and sustaining and was much appreciated by our film crew after a long, dusty day's work.
• Traditionally you would use the flat *reshteh* noodles, but any will work, even small round pasta if you prefer. Gently stir all the ingredients together at the table to produce a silky textured soup-stew.

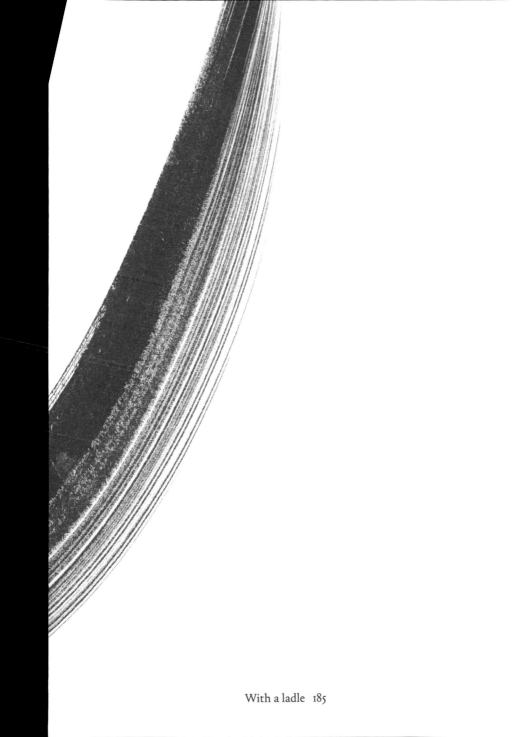

With a ladle 185

ORZO, SMOKED MOZZARELLA, THYME

Pale, soft and deeply cosseting. Notes of thyme, cheese and smoke.

Serves 2

garlic 3 cloves
butter 30g
orzo 200g
vegetable stock 500ml

milk 200ml
thyme 8 bushy sprigs
smoked mozzarella 60g

Peel and thinly slice the garlic. Warm the butter in a deep pan, add the garlic and fry for a couple of minutes till pale gold, then add the orzo, stock and milk and bring to the boil. Pull the leaves from the thyme and add to the pasta together with a half teaspoon of salt, then simmer for fifteen minutes till the orzo is plump, soft and creamy. The sauce should be similar in texture to a risotto.

Grate the smoked mozzarella, then fold into the orzo with a generous grinding of black pepper.

• If I could choose just one bowl of food for a cold night, it would most likely be this. Understated, calm and effortlessly easy, it has a comforting texture and the deep, savoury satisfaction you get with cheese and pasta.
• There are options here. Change the cheese to a fruity, deeply mature Cheddar. Use trofie or strozzapreti pasta in place of the orzo and use smoked garlic for an extra layer of warmth.
• Once the dish is ready, you could also spoon it into a heatproof dish, cover the surface with more grated cheese and grill till bubbling.

TAHINI, SESAME, BUTTERNUT

Sweet and nutty.

Serves 4

butternut squash or pumpkin 1kg
vegetable stock 1 litre
rosemary 3 or 4 sprigs
sesame seeds 3 tablespoons

olive oil 3 tablespoons
chestnuts, canned or
    vacuum-packed 8
tahini 4 tablespoons

Peel and halve the butternut squash, remove the seeds and cut into large chunks, then put into a large saucepan with the vegetable stock and bring to the boil. Cover with a lid and leave to simmer for ten minutes until soft enough to crush.

Ladle the squash and its stock into a blender, process in batches until smooth and return to the pan. Remove the leaves from the rosemary and finely chop. You need enough to fill a tablespoon. Toast the sesame seeds in a dry, shallow pan over a moderate heat until golden, then add the olive oil and the rosemary. Crumble the chestnuts into the pan and cook for a minute or so until all is warm and deeply fragrant.

Bring the soup almost to the boil, checking the seasoning as you go, then ladle into soup bowls. Speckle the soup with a tablespoon of tahini in each bowl, then scatter some of the chestnut and sesame seed seasoning over the surface.

• Some people don't peel butternut squash before using it in a soup. Much depends on the thickness of the skin and the age of the squash. If the skin is thin, then it is fine not to peel it. If you are using a pumpkin, remove the skin. It is important to process the soup in batches rather than all at once, when it is likely to overflow. A stick blender works a treat. *(continued)*

• Use mushrooms instead of the chestnuts. I prefer small brown buttons, sliced in halves or quarters and cooked for a minute or two with the sesame oil and rosemary. A few drops of sesame oil, trickled into the soup as you serve, are worth a thought. I like to eat this soup with thick pieces of toasted sourdough bread, spread with cream cheese.

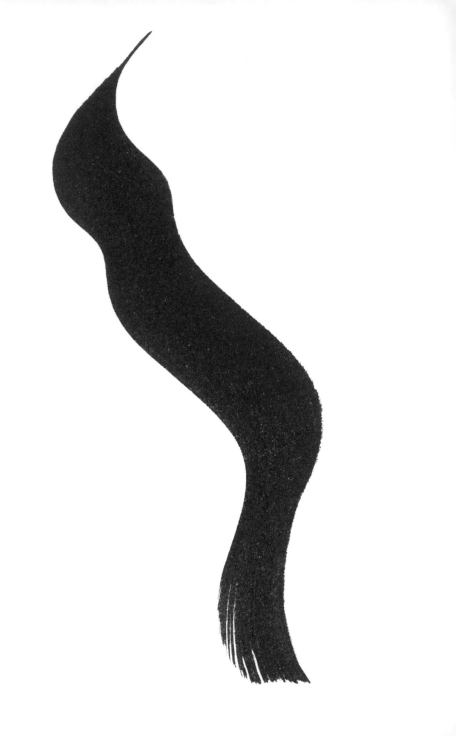

ON THE HOB

As the evenings cool, out come the carbs. The row of jars on the top shelf, the glass ones with bright orange rubber rings, their contents beige, brown, creamy-white or yellow. There is freekeh, the roasted green wheat; fine golden polenta and the brown rices. There is white rice to steam for breakfast and pearl barley to toss with tufts of kale and chalk-white cheese. A frosty night is when tiny beads of fregola get their moment in the moonlight, cooked with blue-black cavolo nero and butter.

The hob, its gas jets on low, is where, for the next few months, we will find pots of lentils ready to be tossed with roast vegetables. It is where fat pearls of mograbia will be cooked in bubbling water, drained and stirred into thick tomato sauces seasoned with cinnamon and cumin. Those gas flames will also fry parsnips for curry and warm spice pastes to fold through bowls of sticky rice.

A deep-sided pan on the hob is where we can simmer milk to replace stock in a silk-textured risotto and warm coconut milk to calm the curry spices for a sauté of roots and spinach. It is where we could boil fregola for tossing with greens and grated Pecorino, and stir thick polenta to calm ourselves after a hectic day.

I like beige food. Suppers of grains and starches that wrap you in the culinary version of cashmere. But there are bright, fresh tastes in the recipes that follow. White rice steamed with lime and green cardamom; a lemon grass sauce for sweet potatoes; tomatoes simmered with chilli and rice vinegar as a sauce for udon noodles. But the real winners are those eaters who need ballast against the cold – a mound of soft yellow polenta with mushrooms and garlic; aubergines simmered in coconut milk; sweet potatoes with fried croquette-like polenta with an accompanying puddle of melting cheese.

Early in the day, as the mist clears, there is porridge to be made and, later, leftover baked suppers that can be reheated the following day on the hob. (A word of advice, let leftovers come up to room temperature before reheating, keep the heat low and the lid on.) I often use the heat of the hob to make a sauce for what is cooking in the oven. A bubble-up of cream, chopped parsley and grain-freckled mustard for a baked potato; a cheese sauce for a butter-roasted onion or melted crème fraîche, lemon juice and chopped tarragon for a sharp accompaniment for a dish of baked parsnips.

Unlike switching on the oven, lighting the hob means supper can be on its way in minutes. I light the hob, get out a deep pan and fry mushrooms till golden, I boil ribbons of pasta till tender then toss the two together with cream, grated Parmesan and chopped herbs and I have the supper of my dreams in minutes. We can't really ask for more than that.

AUBERGINE, TOMATO, COCONUT MILK

Silky aubergines. Soft spice.

Serves 4 with rice

garlic 4 cloves
ginger 40g
red chilli 1 large
small aubergines 700g
groundnut oil 4 tablespoons
ground cumin 4 teaspoons
ground coriander 4 teaspoons

garam masala 4 teaspoons
ground chilli 2 teaspoons
assorted tomatoes 500g
tomato passata 250ml
coconut milk 500ml
coriander leaves a handful

Peel and finely slice the garlic, then peel and finely chop the ginger. Cut the chilli into fine slices. Cut the aubergines in half.

Warm the groundnut oil in a deep pan, add the garlic, ginger and chilli and allow to sizzle for a few seconds, then add the aubergines, cut side down. As the aubergines start to brown add the cumin, coriander, garam masala and ground chilli and continue cooking for two minutes.

Roughly chop and add the tomatoes, let them soften for a couple of minutes, then pour in the passata and bring to the boil. Lower the heat and stir in the coconut milk and a generous grinding of salt. Let the ingredients bubble quietly together for a good ten minutes, until the aubergines are soft enough to crush with a spoon, then serve with a little chopped coriander if you wish.

• Basmati rice is a fine accompaniment to this. Wash the rice in warm water, then cook in a small pan under a tight lid, with enough water to cover the grains by 3cm, for ten minutes. Remove from the heat and leave, still covered by its lid, for ten minutes before fluffing up with a fork. A few cloves, green cardamom pods and a stick of cinnamon would be my preferred seasonings, added at the beginning of cooking.

CARROTS, RICE, CORIANDER

Soft rice. Sweet, earthy flavours.

Serves 4
For the soup:
carrots 400g
coriander, leaves and stems 10g
butter 30g
a large shallot
garlic 2 cloves
olive oil 3 tablespoons

arborio rice 250g
vegetable stock 600ml

To finish:
carrot 100g
olive oil 2 tablespoons
coriander leaves a small handful

Make the soup: Peel and roughly chop the carrots. Bring half a litre of lightly salted water to the boil in a deep pan, add the carrots, then cook for fifteen minutes or so, till tender. Put the carrots and their liquid into a blender with the coriander and butter and process to a smooth purée.

Peel and finely chop the shallot and garlic. Warm the olive oil in a deep-sided frying pan, add the shallot and garlic then let it cook over a moderate flame for about three or four minutes till fragrant, then add the rice. Ladle in the stock, a little at a time, stirring almost constantly. Continue cooking for fifteen to twenty minutes until the stock has been absorbed and the rice is tender but not soft.

Stir the puréed carrots into the rice and simmer for five minutes then add ground black pepper and salt. Peel and very finely dice the remaining carrot. Warm the remaining olive oil in a pan and add the carrot. Cook for three or four minutes till the carrot is approaching tenderness. Finely chop the coriander then spoon the soup into dishes or plates and scatter over the diced carrots and coriander.

FREGOLA, GREENS, PECORINO

The smoky flavours of mushrooms and greens.

Serves 2

dried mushrooms 25g

greens such as cavolo nero 200g

Pecorino 200g

fregola 200g

butter 25g

Bring 1 litre of water to the boil, add the dried mushrooms, remove from the heat and cover with a lid. Set aside for fifteen minutes to infuse.

Wash and finely shred the cavolo nero. Finely grate the Pecorino. Bring the mushroom stock back to the boil, add the fregola and cook for twelve to fifteen minutes until tender but not soft. The pasta should still have a slight chewy quality. Drain the fregola and mushrooms.

Melt the butter in a frying pan and add the shredded greens, frying them for a minute or two till they start to brighten and wilt. Fold in the drained pasta and mushrooms, then the grated Pecorino. Pile into shallow dishes.

• Fregola is easy to find in Italian food stores. I like to look out for the Sardinian toasted variety, which has a pleasing nutty flavour.

• In place of cavolo nero, you could use crinkly-leaved Savoy cabbage or young kale.

• The quantity of Pecorino sounds alarming, but it is there to bring the whole dish together and offer a luxurious contrast to the otherwise humble ingredients.

GNOCCHI, PEAS, EGG YOLK

Hot stock, green peas and the allure of dumplings.

Serves 2

gnocchi 400g

vegetable stock 500ml

frozen peas 350g

egg yolks 2

Parmesan, grated 4 tablespoons

Bring a deep pan of water to the boil and salt it generously. Add the gnocchi and let them cook for three or four minutes, until they come to the surface, or follow the instructions on the packet. Bring the stock to the boil and add the peas.

Remove the gnocchi with a draining spoon and divide between two bowls. Ladle over the stock and peas, then separate the eggs and lower a yolk into each bowl. Sprinkle two tablespoons of grated Parmesan over each serving. Stir the yolk into the stock and peas as you eat, making an impromptu sauce.

HARICOT BEANS, RICE, ONIONS

A dinner of quiet sustenance.

Serves 2

onions, medium 2
olive oil 3 tablespoons
garlic a clove
white long-grain rice 150g
cloves 2
black peppercorns 6

haricot beans 1 × 400g can
butter 60g
cumin seeds 2 teaspoons
ground turmeric 1 teaspoon
chopped dill a handful

Peel the onions and thinly slice into rounds. Warm the olive oil in a saucepan, add the garlic, peeled and finely sliced, then add the onions and leave them to cook over a moderate heat until they are soft and golden. Remove from the heat and take the onions out of the pan but keep the pan for later.

Wash the rice in warm water, drain and transfer to a saucepan, then pour in enough water to cover it by 2cm. Salt the water, add the cloves and the peppercorns, then bring it to the boil. Lower the heat so that the rice simmers and cover the pan tightly with a lid, then leave it for ten minutes. Turn off the heat, leave the lid in place and allow the rice to rest for five minutes.

Drain the beans. Melt the butter in a pan (use the same one you cooked the onions in), then use it to cook the cumin seeds and turmeric for a minute or two until fragrant. Add the drained beans and continue cooking until they are hot.

Run a fork through the rice to separate the grains, check the seasoning and remove the cloves. Stir in the dill, then divide between two bowls. Spoon over the haricot beans and scatter the fried onions on top.

• I suggest haricot beans here, but black-eyed beans, flageolet, butter beans or chickpeas are fine too. What matters is the hearty, cheap-as-chips marriage of beans and rice.

LENTILS, SWEDE, PAPRIKA

Sweet, roasted roots. The heartiness of lentils.

Serves 2

red onion 1	paprika, hot 1 teaspoon
swede 400g	paprika, sweet 1 teaspoon
a large carrot	vegetable stock 500ml
rosemary 4 sprigs	red lentils 100g
thyme 10 small sprigs	small brown lentils 100g
olive oil 5 tablespoons	pappardelle 300g

Set the oven at 200°C/Gas 6. Peel the onion and cut into quarters. Peel the swede and cut into 2cm chunks, then scrub the carrot and cut into similar-sized pieces. Put the onion, swede and carrot into a roasting tin together with the rosemary, thyme and olive oil and turn with a large spoon to evenly coat each piece.

Roast for an hour or until the vegetables are golden brown and tender, turning every twenty minutes or so.

Tip the contents of the roasting tin into a large casserole, add the paprikas, deglazing the roasting tin with the vegetable stock as you go, and add to the casserole along with the rest of the stock. Add the lentils, season thoughtfully, and place on the hob at a low simmer for twenty minutes, stirring from time to time, or until the red lentils have virtually collapsed and the brown lentils are tender.

Cook the pappardelle in deep, generously salted boiling water, for nine minutes or until they are cooked to your liking. Drain and toss with the lentil ragù.

• I am very fond of swede, steamed and mashed with butter and black pepper, and I have used it sliced and cooked with cream and cheese with

(continued)

great success. But swede remains one of those vegetables I find all too easy not to see when I look in the vegetable rack. Every time I cook with swede I ask myself why I don't do so more often.

• Swede responds to most of the spices and especially paprika and cumin. It makes a good mash when stirred through mashed carrots, calming the carrots' overly sweet notes.

On the hob   209

MILK, MUSHROOMS, RICE

As soft and comforting as a cashmere blanket.

Serves 2

milk 800ml	garlic 2 cloves
dried mushrooms 20g	Grana Padano 100g
bay leaves 2	dill 15g
black peppercorns 10	olive oil 4 tablespoons
half an onion	risotto rice 400g
shallots 2	butter 40g

Pour the milk and 100ml of water into a deep saucepan, add the dried mushrooms, bay, peppercorns, onion half (peeled), and bring to the boil. Remove from the heat and set aside, covered with a lid, for about half an hour.

Peel and finely chop the shallots and the garlic. Grate the cheese and finely chop the dill.

Lift the mushrooms from the milk with a draining spoon and finely chop them in a food processor, then return them to the milk. Discard the bay and onion.

Warm the oil in a pan, add the shallot and garlic and cook over a moderate heat till translucent but not coloured. Stir in the rice, then add the mushroom milk, ladle by ladle, only adding more as each ladleful is absorbed. When all but 100ml of the milk has been used, stir in the butter and beat with a wooden spoon for one minute until the rice is thick and creamy, then add the remaining milk.

Stir in the grated cheese, then the dill.

• I admit I was sceptical when James first suggested a milk-based risotto, but one taste and I was convinced. The milk increases the rice's ability to soothe and calm. The bay and onion are essential to provide a mild, savoury back note.

MOGRABIA, TOMATOES, LABNEH

Earthy spice. Fruity sauce. The sustenance of carbs.

Serves 2

onions, medium 2
olive oil 3 tablespoons
mograbia 200g
chopped tomatoes 1 × 400g can

Berbere spice mix 2 teaspoons
cherry tomatoes 200g
tomatoes, medium-sized 150g
labneh 125g

Peel the onions and slice them into 1cm rings. Warm the olive oil in a shallow pan and cook the onions till deep gold and lightly caramelised – a good thirty minutes or more.

Bring a deep pot of water to the boil, rain in the mograbia and boil for about twenty-five minutes till tender.

Put the chopped tomatoes in a deep casserole, add the spice mix (it is already toasted so no need to fry), then bring to the boil, adding the cherry and medium-sized tomatoes, roughly chopped. Simmer for twenty minutes, stirring regularly, then add the drained mograbia, season generously, and continue to cook for a further fifteen minutes until you have a rich, thick sauce.

Serve with the caramelised onions and a spoonful or two of labneh.

• Labneh is available in Middle Eastern stores and many large supermarkets. Should it evade you, use thick, strained yoghurt or spoonfuls of fromage frais.
• The canned chopped tomatoes collapse into the sauce. The roughly chopped fresh tomatoes provide the main texture of the dish, whilst the cherry tomatoes introduce pleasing bursts of sweet-tartness.

MUSHROOMS, SPINACH, RICE

Red spice, white rice. Sticky, spicy, satisfying.

Serves 2

sushi rice 190g
bay leaves 2
black peppercorns 8
roasted salted peanuts 75g
young spinach leaves 100g

Thai red curry paste 50g
100ml groundnut oil plus a
 little extra
king oyster mushrooms 4
coriander leaves a handful

Put the sushi rice in a medium-sized saucepan and pour 300ml of warm water over it. Set aside for half an hour.

Place the pan over a moderate heat and add a half teaspoon of salt, the bay leaves and whole black peppercorns. Bring to the boil, then lower the heat to a simmer and cover with a tight lid. Let the rice cook for twelve minutes, then remove from the heat.

Using a food processor, make a paste of the roasted, salted peanuts, the washed and dried young spinach leaves, the Thai red curry paste and 100ml of groundnut oil.

Cut the king oyster mushrooms into thick slices. Warm a shallow layer of groundnut oil in a frying pan, add the mushrooms and cook till golden, then turn and cook the other side. Lift half the mushrooms from the pan and drain on kitchen paper.

When the rice is ready, add the paste to the mushrooms left in the pan and stir for two or three minutes till the colour darkens slightly. Fold in the cooked rice. Chop a small handful of coriander leaves, fold them into the rice, then serve with the reserved, drained mushrooms.

• The meaty quality of king oyster mushrooms is perfect with the sticky, spicy rice, but you could use large field mushrooms too.
• As it stands this is quite spicy, so stir in more or less curry paste as the fancy takes you.

ORECCHIETTE, CAULIFLOWER, CHEESE

The reassurance of pasta. The soothing notes of cheese and cauliflower.

Serves 2

Parmesan 200g

a medium cauliflower

butter 30g

olive oil 3 tablespoons

orecchiette 200g

double cream 250ml

dill fronds 10g

Put a deep pan of water on to boil and salt it generously. Finely grate the Parmesan. Cut the cauliflower florets from the main stalk.

Warm the butter and oil together in a shallow pan, then fry the cauliflower florets for five minutes or so until lightly crisp and golden. Put the orecchiette into the boiling water and cook for about nine minutes or until tender.

Pour the cream into the cauliflower, add the grated Parmesan and lower the heat. Drain the pasta and add it to the cauliflower and cream. Season with black pepper. Chop the dill fronds, stir into the pasta and serve.

• Broccoli, Brussels tops, Brussels sprouts and kale are all perfectly acceptable in place of the cauliflower. The pasta is interchangeable too, but it's worth choosing one that will hold a little puddle of sauce, such as casarecce, anelli, cavatelli or ditalini. I suggest you avoid the larger varieties of penne, which is rather like eating pieces of rubber tubing.

PARSNIPS, CASHEWS, SPICES

Sweet roots. Soft spice. Crunchy nuts.

Serves 2

parsnips 500g
onions, medium 2
garlic 2 cloves
groundnut oil 4 tablespoons
ground turmeric 2 teaspoons
ground cumin 2 teaspoons

ground coriander 2 teaspoons
curry powder 2 teaspoons
coconut milk 1 × 400g can
young spinach 200g
roasted cashews 75g
a red chilli, mild

Peel the parsnips and cut them into large pieces. Peel and roughly chop the onions. Peel and finely chop the garlic.

Warm the oil in a large, deep saucepan, add the parsnips and onions and sauté for about fifteen to twenty minutes till lightly coloured, stirring regularly, then add the garlic. Fry briefly, then sprinkle in the ground spices and curry powder and continue cooking for a minute or two till fragrant. Pour in the coconut milk, bring to the boil then lower the heat.

Wash the spinach, then put the still-wet leaves into a saucepan and cover tightly with a lid. Cook over a high heat for a minute or two till the leaves wilt. Remove from the pan and squeeze dry. In a blender, process the spinach leaves with the cashews and the chilli and stir into the curry. Check the seasoning and serve.

• Much earthy sweetness here. The chilli adds a kick but not so much as to mask the dish's soothing qualities.
• I do think rice is called for here and quite plain rice too. Salt, lemon zest and maybe some chopped coriander is all the seasoning it requires.

PEARL BARLEY, KALE, GOAT'S CHEESE

Nutty grains, melting cheese, tender greens.

Serves 2

vegetable stock 800ml	curly kale 150g
pearl barley 200g	olive oil 4 tablespoons
smoked garlic 1 head	goat's cheese 200g

Heat the stock in a deep pan and tip in the pearl barley. Cut the smoked garlic in half horizontally, slicing through the skin and cloves, drop into the stock and simmer for 35 minutes till the barley is tender.

Cut the stems from the kale, setting the leaves aside. Roughly chop the stems. Pile the leaves on top of each other and finely shred into ribbons. Remove the smoked garlic, scoop out the flesh with a knife and crush to a paste. Discard the skins. (Any cloves that have fallen in the barley during cooking can be left in.)

Heat the olive oil in a large, shallow pan, add the chopped kale stems and cook for a few minutes till tender and bright. Stir in the crushed smoked garlic, then add the shredded kale leaves. Sizzle for a couple of minutes then fold into the pearl barley, together with the crumbled goat's cheese.

• I have suggested kale because of its stridency against the soft, smoky grain, but almost any brassica is applicable here.
• Mozzarella would add strings of cheesy joy to the barley, as would Fontina.
• I like the nutty quality of pearl barley but this recipe could also be made with orzo pasta for a softer consistency.

POLENTA, GARLIC, MUSHROOMS

Soupy, starchy, nannying. Food to soothe and strengthen.

Serves 2–3

vegetable stock 1 litre
coarse polenta (bramata) 150g
butter 80g
double cream 250g
grated Parmesan 100g

For the mushrooms:

large field mushrooms 2
chestnut mushrooms 150g
olive oil 5 tablespoons
butter 80g
garlic 3 cloves
parsley, chopped 2 heaped
  tablespoons
chopped sage leaves 4

Bring the stock to the boil in a deep-sided, heavy-based saucepan. As the stock starts to boil, rain in the polenta, stirring all the time with a wooden spoon. Continue stirring, making sure to get deep into the corners of the pan. Continue cooking, turning the heat down a little, for forty minutes, stirring almost constantly.

Cut all the mushrooms into slices about 1cm thick. Warm the oil in a large frying pan, then add 80g of butter. When the oil starts to fizz, add the mushrooms. You may find it easier to do this in two batches. Cook the mushrooms until soft and honey-coloured, turning them once during cooking, then remove them to a dish using kitchen tongs. Don't wash the frying pan just yet.

Stir the remaining 80g of butter, the cream and the Parmesan into the polenta, then pour in enough boiling water to give a soft texture that will fall easily from the spoon. Peel and finely slice the garlic, then add to the frying pan used for the mushrooms, letting it brown lightly in the leftover butter. Return the mushrooms briefly to the pan, add the chopped parsley and sage and make sure everything is hot. *(continued)*

Spoon the polenta into a serving dish, add the mushrooms and garlic and serve.

• Of the two main sorts of polenta, I like to use the coarse (bramata) here. The finer version will work, but the consistency is less satisfying and produces a thinner result.
• You can omit the sage leaves if you wish – it is hardly the most useful herb to have in the house – but I often add just a few of them to polenta and, sometimes, to mashed potato.

On the hob   225

POLENTA, THYME, TALEGGIO

Crunchy carbs. Melting Taleggio.

Enough for 3–4

coarse polenta (bramata) 125g
thyme leaves 1 teaspoon
oil, for deep frying

crème fraîche 4 tablespoons
Taleggio 150g

Bring 750ml of water to the boil in a heavy-based, deep-sided pot and salt generously. Add the polenta in a continuous stream, then lower the heat and stir regularly for thirty minutes until you have a thick porridge. Stir in the thyme. Scoop the polenta into a loaf tin, approximately 20 × 12cm, lined with clingfilm, smooth the surface and wrap the film over the top. Leave for an hour or so to cool.

Unwrap the polenta and turn out onto a chopping board or plate. Break into about twenty-four pieces, each one about 30–35g in weight. The rougher the break the more interesting the texture will be when they are fried.

Heat the oil to 180°C in a deep pan. Lower the pieces of polenta into the hot oil, one or two at a time, taking care not to overcrowd the pan. Let the polenta cook for five or six minutes, turning from time to time, then removing with a draining spoon as soon as they are crisp and golden. Place on kitchen paper.

Warm the crème fraîche in a heatproof bowl over a pan of simmering water. Add the cheese (cut into small pieces) and let it melt. Stir gently.

Serve the crisp pieces of polenta with the Taleggio cream.

• A crisp salad on the side. Curls of frisée crisped in iced water; white chicory and watercress would be my preference.

RICE, BROCCOLI, PAK CHOI

The calmness of rice. The vibrancy of greens.

Serves 2

white basmati rice 125g	Tenderstem broccoli 200g
cloves 4	olive oil 2 tablespoons
bay leaves 2	five-spice powder 1 teaspoon
green cardamom pods 6	nigella seeds 2 teaspoons
whole black peppercorns 8	toasted sesame oil 2 teaspoons
pak choi 100g	soy sauce to taste
spring onions 3	

Put the basmati in a mixing bowl, cover with warm water and run your fingers through it, until the water becomes cloudy. Repeat twice, by which time the water should be almost clear. Put the rice into a small lidded saucepan, pour in enough water to cover it by 3cm, then add the cloves, bay, green cardamoms lightly cracked, the whole peppercorns and a half teaspoon of salt.

Bring to the boil, then cover tightly with a lid and lower the heat so the steam barely lifts the lid. Leave to cook for ten minutes, then take off the heat and set aside without removing the lid.

Finely shred the pak choi and the spring onions. Cut the broccoli into small pieces. Warm the oil in a shallow pan that doesn't stick and add the pak choi, spring onions and the Tenderstem, frying till bright green and tender. Stir in the five-spice powder, nigella and the sesame oil then fluff the rice up with a fork and fold into the greens.

Spoon the rice and greens onto plates or dishes, then trickle with a little soy sauce.

• There is an almost endless list of greens you could use in place of the broccoli, such as shredded Brussels sprouts, cavolo nero, Savoy cabbage or kale.
• This is a useful way to use leftover steamed rice. Heat the rice gently and thoroughly in a wide frying pan.

RICE, LEMON, LIME

Warm rice. A whiff of citrus.

Enough for 4 as an accompaniment

white basmati rice 250g	lime leaves 6
green cardamom pods 12	a lime
black peppercorns 10	a lemon
bay leaves 2	a small orange
cloves 4	

Tip the rice into a large, wide bowl, cover with warm water and move the grains around with your hand until the water turns milky. Drain off the water, then fill the bowl with more warm water and repeat. Do this a total of three times or until the water is almost clear. Tip the drained rice into a medium-sized saucepan.

Lightly crack the cardamom pods and add to the rice together with the whole black peppercorns, the bay leaves, cloves and lime leaves. Grind in a half teaspoon of sea salt and pour in enough water to cover the grains by 2cm. Bring the water to the boil, then lower the heat and cover with a tight lid. Allow the rice to putter away for ten minutes, then remove from the heat and leave for ten minutes, lid in place.

Finely grate the zest from the lime, lemon and orange, making sure not to include any white pith. Remove the lid from the rice, run the tines of a fork through the grains to separate them, then fold in the grated citrus zest and a generous grinding of black pepper.

• Sometimes, all that is required is a bowl of plain rice. This is often lunch for me, with a cup of miso soup. I should mention that it also makes a pleasing accompaniment to almost anything.

SWEET POTATO, COCONUT MILK, CASHEWS

Sweet and milky, with deep spice notes. Food to warm the soul.

Serves 4 with roti or rice

sweet potatoes 600g
small, waxy potatoes 400g
sunflower oil 2 tablespoons
green cardamom pods 15
cumin seeds 2 teaspoons
ground cinnamon 1 teaspoon
ground coriander 2 teaspoons
garlic 4 cloves

small, hot red chillies 4
lemon grass 2 stalks
ginger 35g
a lime
groundnut oil 2 tablespoons
cashews 75g
coconut milk 400ml
coriander leaves a handful

Peel the sweet potatoes and wipe the waxy ones, then cut each into 2cm dice. Warm the oil in a shallow pan, then lightly brown the vegetables. They should be tender.

Crack the cardamom pods and remove the seeds, then put them in a spice mill or coffee grinder with the cumin, ground cinnamon and ground coriander and grind everything to a powder.

Peel the garlic, roughly chop the chillies, finely slice the lemon grass and peel the ginger and grate coarsely. Finely grate the zest from the lime and squeeze the juice. Put all into the bowl of a food processor together with the ground spices, groundnut oil and cashews. Process to a thick paste.

Put four tablespoons of the paste in a non-stick frying pan and cook over a moderate heat for two or three minutes till fragrant and slightly darker in colour. Stir constantly to prevent it sticking. Stir in the coconut milk and coriander leaves and simmer for three to four minutes, stirring almost constantly.

Divide the vegetables between four plates and spoon over some of the sauce. *(continued)*

• This is a massaman-style curry, with the crunch of cashews. I marginally prefer soft, warm naan to rice to scoop this up, but either will do the trick. I really think it is worth warming the bread; in fact I would suggest it is essential. Cold naan is a friend to no one.

• Curries such as this can be made with other vegetables too. I particularly like to use plump chestnut mushrooms cut into quarters before sautéing, perhaps mixed with halved Brussels sprouts.

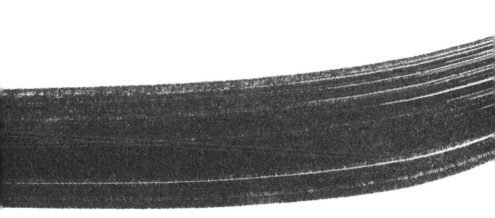

TAGLIATELLE, DILL, MUSHROOMS

Autumn mushrooms, ribbons of pasta, a breath of aniseed.

Serves 2

dill 25g
grated Parmesan 125g
double cream 250g
king oyster mushrooms 200g
brown Shimeji or other small
   mushrooms 150g
chanterelles or other 'wild'
   mushrooms 30g
butter 30g
olive oil 3 tablespoons
tagliatelle or fettucine 250g

In a blender or food processor, reduce the dill and Parmesan to fine crumbs. Tip into a saucepan, add the cream and bring gently to the boil, stirring until the cheese melts. Cover with a lid, remove from the heat and set aside.

Slice the oyster mushrooms into pieces the thickness of a pound coin. Clean and trim the other mushrooms. Melt the butter in the oil in a shallow pan over a moderate heat. Sauté the king oyster mushrooms for three or four minutes till tender, then add the remaining mushrooms and cook for three minutes or so.

Bring a deep pan of water to the boil, salt generously, then add the pasta and let it cook till al dente. Drain, return to the pan, then fold in the cream, herb and cheese mixture, followed by the cooked mushrooms. Divide between plates.

• Use whatever mushrooms you have to hand. Chestnut, even little brown buttons, will work well enough.

TOMATO, CHILLIES, UDON

Fruity, frugal and a little fiery.

Serves 2

cherry tomatoes 500g
garlic 4 cloves
a large red chilli
groundnut oil 3 tablespoons

udon noodles 250g
rice vinegar 1 tablespoon
mirin 2 teaspoons
coriander leaves a handful

Heat the overhead (or oven) grill to high. Halve the cherry tomatoes, peel and thinly slice the garlic, cut the chilli into very thin rounds. Put the tomatoes, garlic and chilli into a shallow roasting tin, then toss with the groundnut oil. Let the tomatoes cook under the grill till some of the skins have blackened.

Bring a deep pan of water to the boil and salt it generously. Lower the noodles into the water and let them boil for three or four minutes, or according to the instructions on the packet. Drain the noodles as soon as they are tender.

Crush the tomatoes with a fork or vegetable masher, then stir in the rice vinegar, mirin and coriander, toss with the noodles and serve in shallow bowls.

• Letting the tomato skins blacken under the grill lends a smoky note to the sauce.
• I sometimes stir a little chilli paste into this too, usually the tearfully hot Korean gochujang.

PUDDING

Sometimes we need pudding. A crisp crusted tart, a slice of warm cake or crumble-topped fruit.

Pudding can be as straightforward as a disc of shortbread, its surface studded with crushed hazelnuts, spread generously with cream and sliced blood oranges. You could finish your dinner with meringue, crushed and stirred into whipped cream, the white depths trickled with bubbling mincemeat or a sauce of hot fruit preserves.

Crumbles can be made with the usual butter, flour and sugar crust or with breadcrumbs, butter and syrup. That crust can shelter apples or damsons, and can take ground nuts or chips of dark chocolate. It can be served hot from the oven, or chilled for breakfast. (There is actually very little difference between muesli with stewed apple and an Apple Betty.)

Some like to end their meal with chocolate. In which case a dish of chocolate pudding, almost liquid in texture, to eat with a teaspoon or scoop from the dish with crisp Italian biscuits may be just the thing. Others may go for broke with clusters of dried fruit held together with dark chocolate and ground pistachios.

Most of my favourite sweet treats are fruit-based. Apples are roasted and crushed, then stirred through with toasted rye crumbs and dried fruit. Bananas baked with maple syrup and puff pastry. A salad of papaya marinated in the juice of passion fruits.

There are a few that could be filed under cheating. (Whatever that means.) In particular, the idea of using ready-made cake in the style of a sponge pudding, served with a jug of cream or, better still, a 'custard' of cream and crushed cardamoms.

I will eat ice cream whatever the weather, but once the evenings draw in I am likely to serve it with melted chocolate or a glass of sherry. Then there are the confections – meringue with cream, crushed biscuits or nougat or macaroons to scatter over the frost-cold surface. Whatever the weather, there must almost always be pudding.

APPLES, CINNAMON, PUFF PASTRY

Warm spice. Crisp pastry.

Makes 12

sharp 'cooking' apples 750g	puff pastry 325g
mixed spice 2 teaspoons	a little beaten egg
ground cinnamon 1 teaspoon	a little ginger syrup or honey
marzipan 250g	

Set the oven at 220°C/Gas 8. Line a baking sheet with parchment.

Peel and core the apples, then roughly chop half of them and put them in a deep saucepan with a splash of water. Bring to the boil, lower the heat and let them simmer until they are soft, frothy and can be crushed with a fork. Remove from the heat.

Cut the remaining apples into small cubes and fold them into the crushed apple with the mixed spice and ground cinnamon. This will give a nice contrast of textures.

Roll the pastry out to a rectangle measuring 36 × 23cm. Turn the pastry so the longest edge is facing you. Cut the marzipan into tiny cubes. Spread the apple over the pastry, leaving a 3cm rim on the left- and right-hand edges, then scatter the marzipan over.

Brush the bare pastry rim with a little of the beaten egg. Roll the pastry up from right to left, press the egg-brushed edges firmly to seal, then cut the roll of pastry into twelve equal slices. Place the slices cut side up on the parchment-lined baking sheet.

Bake in the preheated oven for fifteen minutes, then remove from the oven and brush with a little ginger syrup or honey.

APPLES, GINGER CAKE, CUSTARD

Wobbling custard. Soft, spiced cake.

Serves 4–5

sweet apples 350g
butter 40g
plain ginger cake 350g
a little demerara sugar

For the custard:

eggs 4
double cream 500ml
full-cream milk 125ml
golden caster sugar 50g
ground cinnamon a pinch
nutmeg for grating

You will also need a baking dish approximately 22–24cm in diameter.
Set the oven at 180°C/Gas 4.

Quarter the apples, remove their stalks and cores, then slice each quarter into three. Melt the butter in a shallow pan over a moderate heat, then add the apples and let them cook for eight to ten minutes till they start to soften. Turn them over carefully with a palette knife and continue cooking till soft.

Make the custard: break the eggs into a mixing bowl and beat with a small whisk to mix whites and yolks. Pour in the cream and milk, add the sugar, ground cinnamon and a little grated nutmeg, combine thoroughly, then pour into the dish.

Crumble the ginger cake into large lumps and scatter over the custard. Add the apple slices, then sprinkle lightly with demerara sugar. Bake for thirty-five to forty minutes till the custard is lightly set. It should wobble a little as you shake the dish. Leave to settle for fifteen minutes. Serve warm, though I should add it is good thoroughly chilled, too.

APPLES, OATS, CHOCOLATE

Deep, tender apple. Dark chocolate. Crisp oats.

Serves 4

sharp 'cooking' apples 1kg
butter 30g

For the crust:
soft, fresh white breadcrumbs 60g
jumbo oats 60g

dark chocolate chips 70g
butter 70g
maple syrup 2 tablespoons
soft, light brown sugar
   2 tablespoons

You will also need an oven dish approximately 22 × 15cm.

Set the oven at 190°C/Gas 5. Peel, core and roughly chop the apples, then put them into a saucepan with 3 tablespoons of water and the 30g of butter and cook over a moderate heat for ten minutes or so, till the apples are turning to fluff. Transfer to the oven dish.

Mix together the soft breadcrumbs, oats and chocolate chips. Melt the butter and maple syrup together in a small saucepan. Scatter the crumbs over the apples, then pour over the syrup and butter, soaking the crumbs as you go.

Sprinkle the sugar on top and bake for twenty-five to thirty minutes till crisp and light gold in colour.

APPLES, PEARS, RYE

Soft apples. Crisp crumbs.

Serves 4–6

sweet apples, large 4
pears, large 3
cloves 6
ground cinnamon 2 teaspoons
an orange
caster sugar 4 tablespoons
a lemon
butter 50g

For the crust:
dark rye bread 100g
golden sultanas 6 tablespoons
mixed spice 2 teaspoons
butter 50g
fresh cranberries 2 tablespoons

Set the oven at 200°C/Gas 6. Peel two of the apples, core them and cut them into thick segments, then do the same with two of the pears. Put the peeled apples and pears in a baking tray or roasting tin and add the cloves and a light dusting of the cinnamon. Remove six strips of orange peel with a vegetable peeler and tuck amongst the fruit.

Cut the remaining fruit into segments, coring them as you go but without peeling them, then put them into a second, smaller tin. Scatter both roasting tins of fruit with the four tablespoons of caster sugar. Halve the lemon and squeeze its juice over, then dot the 50g of butter amongst the fruit. Bake both tins for about forty-five to fifty minutes until all is soft.

Crumble the dark rye bread into a bowl and stir in the sultanas and mixed spice. Melt the butter in a small pan, add the crumbled rye bread and let it toast over a moderate heat till lightly crisp. Remove and set aside.

Take the tray of peeled fruit from the oven. Remove the peel and spices, then crush the fruit with a vegetable masher and place in a serving dish.

Scatter the cranberries over the pears and apples in the oven. When the fruit is golden and sticky and the cranberries are starting to burst, remove them from the oven. Spoon the roast fruit over the purée, then scatter with the toasted rye bread.

BANANAS, BUTTER, PUFF PASTRY

Crisp pastry. Warm banana. The scent of maple syrup.

Makes 4

ready-rolled all-butter puff
    pastry 125g
butter 40g

bananas 2
icing sugar 2 tablespoons
maple syrup 4 tablespoons

You will need two baking sheets.

Set the oven at 220°C/Gas 8. Place one of the baking sheets upside down in the oven and line the other with baking parchment.

Roll the pastry thinly into a rectangle large enough from which to cut four discs of pastry 12cm in diameter. Using a large cookie cutter or small plate as a template, cut out the four discs and transfer them to the parchment-lined baking sheet. Melt the butter in a small saucepan and slice the banana into pieces as thick as a pound coin.

Divide the banana slices between the pieces of pastry, placing them, overlapping, in the centre of the discs of pastry. Brush the bananas and pastry with melted butter, then sift the icing sugar over them.

Place the baking sheet on top of the hot sheet already in the oven and bake for about eighteen minutes until the pastry is golden. Spoon the maple syrup over the tarts and return to the oven for five minutes. Remove from the baking sheet with a palette knife or fish slice.

• Very finely sliced apples will work in place of the bananas, though they do need to be thoroughly basted with butter before baking. My favourite for this treatment is fresh apricots (they need to be exceptionally ripe), and I should perhaps mention that the canned variety, for which I retain a certain affection, are perfect candidates for a spot of high-speed baking.

(continued)

• Little tarts such as these benefit from singed edges. The contrast of the almost burned pastry with the sweet fruit and fridge-cold cream is a delight in the way they are in a tarte Tatin. That said, it is worth keeping a close eye on them during their time in the oven. Once the pastry is brushed with syrup and returned to the oven it will be ready in the blink of an eye.

• The sheets of ready-rolled all-butter puff pastry are the ones to look out for. The thick blocks take an age to thaw and are generally too large unless you are making tarts for a party.

CHOCOLATE, DRIED FRUIT, PISTACHIOS

Chewy chocolate clusters.

Makes 6

dark chocolate 100g
soft dried prunes 60g
dried apricots 60g
skinned hazelnuts 40g
golden sultanas 40g

For the pistachio sugar:

shelled pistachios 15g
mint leaves 10
caster sugar 10g

Line a baking sheet with parchment. Break the chocolate into a heatproof bowl, place over a small saucepan of simmering water and leave to melt. Try to avoid stirring, just occasionally push any unmelted pieces into the liquid chocolate. Remove the bowl from the heat.

Roughly chop the prunes and apricots. I like to slice them into small strips, each slightly thicker than a match. Roughly chop the hazelnuts, then add to the chocolate with the apricots, prunes and sultanas and stir lightly.

Place large spoonfuls of the chocolate-coated fruit and nuts on the parchment-lined baking sheet.

Put the pistachios, mint leaves and sugar into the bowl of a food processor and blitz to a fine and pale green powder. Sprinkle over the chocolate clusters and leave in the fridge until set.

CHOCOLATE, DULCE DE LECHE, CANTUCCI

Fragile crust. Oozing chocolate.

Serves 2

dark chocolate 100g

dulce de leche 2 tablespoons

eggs 2

caster sugar 100g

cantucci, to serve

Set the oven at 160°C/Gas 3. Break the chocolate into small pieces and leave to melt, without stirring, in a heatproof bowl suspended over a pan of simmering water. As the chocolate melts, gently stir in the dulce de leche and turn off the heat.

Break the eggs into a large bowl, add the sugar and beat until thick and fluffy. Doing this in a food mixer with a whisk attachment will give the best results. Stir the chocolate and dulce de leche into the mixture.

Transfer to two ramekins, using a rubber spatula. Put the ramekins into a roasting tin or baking dish. Pour enough boiling water to come halfway up the sides of the ramekins, then bake for twenty minutes until the surface is lightly crisp and the inside thick and creamy.

Serve with a teaspoon and, if you wish, cantucci biscuits.

• You need only two three or three stirs to incorporate the dulce de leche into the chocolate. Any more and you might over-mix it.
• A soft crust appears on these puddings as you bake them, whilst the inside stays rich and fondant-like. It will stay like that for an hour or two, should you wish to make them a little ahead of time.
• Heatproof china ramekins are ideal for these, but you can bake them in ovenproof cups too, or even metal tali dishes.
• I have baked these chocolate puddings and eaten them the following day, when they are like thick, fudgy chocolate mousse.

CHOCOLATE, ORANGE ZEST, CANDIED PEEL

Warm scones, nuggets of chocolate, a spritz of orange.

Makes 12 small scones

plain flour 500g

baking powder 3 teaspoons

chilled butter 75g

caster sugar 1 tablespoon

an orange (medium-sized)

candied orange peel 75g

dark chocolate 75g

milk 300ml

a beaten egg

Set the oven at 220°C/Gas 8. Line a baking sheet with parchment.

Sift the flour and baking powder together into a large bowl. Cut the butter into small cubes, then rub into the flour with your fingertips until you have what looks like soft, fresh breadcrumbs. Stir in the sugar. Finely grate the zest from the orange into the flour, taking care not to include any of the bitter white pith underneath.

Chop the candied peel into tiny, gravel-sized pieces and add to the bowl. Chop the chocolate into small pieces, a little larger than the cubes of peel, and stir in, together with the milk, and mix to a soft, but rollable dough.

Turn out onto a lightly floured board and pat into a thick round. Using a 6cm biscuit cutter, cut out twelve scones and place them, a little way apart, on the baking sheet. Brush the top, but not the sides, of the scones with a little of the beaten egg and bake for 10–12 minutes till golden. Transfer the scones to a cooling rack and serve when still slightly warm.

• The best candied peel is that which you buy in the piece from Italian delicatessens and food halls. It is by far more juicy and sweet-sharp than the little nuggets of 'candied peel' you buy ready-chopped.

• Despite the chocolate, I do think these benefit from being spread with butter as you would any scone. But no cream and jam. A splodge of marmalade might be fun though.

DAMSON, ALMOND, SUNFLOWER SEEDS

Sweet homely crust. Juicy fruit.

Serves 4-6

damsons 750g
caster sugar 2 tablespoons

For the crumble:
butter 90g

plain flour 75g
ground almonds 75g
caster sugar 40g
sunflower seeds 3 tablespoons
cream, to serve

Set the oven at 180°C/Gas 4. Wash the damsons, then put them, still wet, into a baking dish approximately 20 × 22cm. Sprinkle them lightly with the two tablespoons of sugar.

Cut the butter into small cubes, then rub into the flour and ground almonds with your fingertips, until it resembles fine crumbs. Alternatively, do this using the food processor, in which case it will be done in seconds. Stir in the 40g of caster sugar and the sunflower seeds.

Shake a few drops of water over the crumble, then shake the bowl so some of the crumbs stick together in larger pieces. You get a more interesting crust that way.

Pile the crust on top of the damsons and bake for about thirty minutes until the fruit is bubbling around the edge of the crust. Serve with cream.

GINGER CAKE, CARDAMOM, MAPLE SYRUP

Hot cake, spice cream, sweet syrup.

Serves 2

double cream 250ml

cardamom pods 8

caster sugar 1 tablespoon

ginger cake 150g

butter 30g

maple syrup 2 tablespoons

Put the cream in a small saucepan. Break open the cardamom pods and crush the seeds to powder in a spice mill or a pestle and mortar. Stir the powder into the cream, add the sugar, then simmer for three minutes and remove from the heat.

Cut the ginger cake into four slices 2cm thick. Warm the butter in a frying pan, add the ginger cake and cook for a couple of minutes until thoroughly hot. Remove and divide between two dishes. Pour the cardamom cream over the ginger cake, then trickle over the maple syrup.

• I say ginger cake, but to be honest any open-textured cake would do. A lemon cake is good here, as are spice cakes and those made with golden syrup. Madeira cake is a bit too dry and tight-crumbed.

• Cardamom has an affinity with sweet baking and is I think perfect here, but you could use a little cinnamon instead if you like. You will need barely a half teaspoon.

• If maple syrup evades you, the syrup from a jar of preserved ginger will work too, bringing a little extra warmth as well as sweetness.

HAZELNUTS, BLOOD ORANGES, RICOTTA

The warmth of nuts and honey. The brightness of oranges.

Serves 6

For the base:
butter 150g
icing sugar 60g
ground almonds 80g
plain flour 75g
skinned hazelnuts 50g
pine kernels 40g
poppy seeds 10g
sesame seeds 10g
flaked almonds 30g
shelled pistachios 40g
small lemon 1

blood oranges 4
double cream 150ml
ricotta 200g

For the syrup:
honey 3 tablespoons
thyme leaves 2 teaspoons

Set the oven at 160°C/Gas 3. Using a food mixer with a flat beater, cream the butter and icing sugar together. Fold in the ground almonds and plain flour. Chop the hazelnuts in half, then mix with the pine kernels, poppy and sesame seeds, flaked almonds and pistachios. Finely grate the zest from the lemon, then fold, together with the seeds and nuts, into the shortbread dough.

Bring the mixture together with floured hands, put on a parchment-lined baking sheet, then lightly pat into a rectangle roughly 22 × 24cm. Bake for twenty-five to thirty minutes until pale biscuit-coloured. Remove from the oven and leave to cool for ten minutes, then slide carefully on to a cooling rack and leave until cold.

(continued)

Remove the peel from three of the oranges with a sharp knife, then slice the fruit thinly and place in a small bowl. Put the cream in a mixing bowl and beat until thick, but not so firm it will stand in peaks. Fold in the ricotta and refrigerate.

Put the juice of the fourth orange, the honey and thyme leaves into a small pan and warm over a low heat until the honey has melted. Slide the shortbread on to a serving plate or board, spoon the chilled ricotta cream over the surface, then cover with the sliced oranges. Spoon over the orange and thyme syrup and serve.

Pudding   269

ICE CREAM, CHOCOLATE, SHERRY

Cold ice cream. Hot chocolate sauce. The deep, sticky sweetness of sherry.

Serves 2

flaked almonds 1 tablespoon

dark chocolate 75g

vanilla ice cream 4 balls

Pedro Ximénez sherry 2 tablespoons

crystallised orange

Toast the flaked almonds in a dry, shallow pan over a moderate heat till patchily golden. Take great care not to get distracted – nuts burn in seconds.

Break the chocolate into a small heatproof bowl suspended over simmering water and leave to melt.

Place the balls of ice cream in two bowls, pour in the Pedro Ximénez, then spoon in the melted chocolate. Finish with the flaked almonds and a tiny slice of crystallised orange and eat immediately.

CRANBERRY, MINCEMEAT, MERINGUE

Crisp meringue. Whipped cream. Sweet mincemeat. Sour cranberries.

Serves 2

cranberries, fresh 125g	meringues 80g
cranberry jelly 150g	mincemeat 200g
double cream 250ml	Marsala 40ml

Put the cranberries in a saucepan with the cranberry jelly and leave over a moderate heat for five to seven minutes till the berries burst.

Pour the cream into a chilled bowl and whisk till thick. The cream should be just stiff enough to sit in soft folds. Crumble the meringues into the cream.

Warm the mincemeat and Marsala together in a small pan until they are hot and bubbling, then remove from the heat.

Divide the meringue and cream between two bowls, then spoon over the hot mincemeat followed by the hot cranberry sauce.

• You could successfully add slices of blood orange to this, too.

PAPAYA, PASSION FRUIT, MINT

Tart, bright flavours for a grey day.

Serves 4

passion fruit 6	persimmon 1
orange 1, small	papayas 2
cardamom pods 4	mint leaves 6

Slice the passion fruit in half, then squeeze their juice and seedy flesh into a small sieve over a bowl. Press the juice and pulp through the sieve with the back of a teaspoon, then discard the seeds. Cut the orange in half and squeeze its juice into the bowl of passion fruit juice.

Crack open the cardamom pods with a heavy weight, such as a mortar or rolling pin, then drop them, whole, into the juice. Cover and place in the fridge.

Slice the persimmon thinly, then add to the passion fruit juice. Slice the papayas in half, scrape out the seeds with a teaspoon and discard them, then remove the yellow skin from each papaya half using a vegetable peeler. Slice the papaya into thick pieces, about the width of a pencil. Mix the papaya pieces with the persimmon and passion fruit juice, tossing the fruits tenderly together.

Cover and leave for a couple of hours, or overnight, in the fridge. Add the mint leaves and remove the cardamom pods - they have done their work - then divide the fruit between four small bowls.

PEARS, RED WINE, PECORINO

Fruity, long-matured cheese, spiced fruits.

Serves 4

pears 4	cloves 4
star anise 2	a cinnamon stick
cardamom pods 10	a bottle of light, fruity red wine
a vanilla pod	aged Pecorino 350g

Peel the pears and tuck them snugly into a stainless-steel saucepan. Add the star anise, cardamom pods – lightly cracked open – the vanilla pod (split in half lengthways) and the cloves. Break the cinnamon stick in half, add to the pan then pour in the red wine.

Bring the wine almost to the boil, lower the heat to a simmer, then leave to cook until the pears are tender. Expect this to take anything from twenty minutes to fifty, depending on the ripeness of your pears.

Slice the cheese, drain the pears, discarding the spices, and put them both on a serving plate.

• You can use the spiced red wine. Once you have removed the pears, turn the heat up, add 100g of caster sugar and let the wine bubble to a deep, garnet-red syrup. Keep your eye on it, but it will probably take a full ten minutes. When all is deep red and glossy, pour into a small jug and leave to cool. Pour over vanilla ice cream.

PUDDING RICE, ROSEWATER, APRICOTS

Creamy rice. A hint of rosewater.

Serves 4
pudding rice 150g
double cream 250ml
full-fat milk 250ml
golden caster sugar 2 tablespoons
a vanilla pod
rosewater 2 teaspoons
orange blossom water 2 teaspoons
pistachios, roughly chopped 50g

rose petals, fresh or dried
  2 tablespoons

For the apricots:
dried apricots 16
half a cinnamon stick
a lemon

Put the rice in a medium, heavy-based pan, then add the cream, milk, sugar and vanilla pod. Bring to the boil over a medium heat, then turn down the heat until the milk is bubbling gently. Allow to simmer for twenty-five minutes or until tender, giving it the occasional stir.

Put 300ml of water in a medium-sized pan, add the dried apricots, cinnamon stick and the juice and shell of the lemon. Bring to the boil, then lower the heat and simmer for ten minutes.

When the rice is soft, remove the vanilla pod and stir in the rose- and orange blossom water. Divide the rice pudding between four bowls. Spoon four apricots over each, with a little of the juice from the pan. Scatter over the pistachios and a few rose petals before serving.

SHERRY, BLOOD ORANGES, CREAM

The nutty warmth of sherry. The vibrancy of oranges.

Serves 4

blood orange 1 small
lemon 1 small
oloroso sherry 75ml
caster sugar 75g
double cream 300ml

To finish:
dark chocolate 50g
candied orange zest 75g
blood oranges 2

Finely grate the zest of one blood orange and the lemon into a mixing bowl. Squeeze the fruits' juice into the bowl, pour in the sherry and stir in the sugar. Mix until the sugar has dissolved, then set aside for a good two hours or even overnight.

Make the chocolate-dipped peel. Melt the chocolate in a small heatproof bowl over a pan of simmering water, removing it from the heat as soon as the chocolate is liquid. Cut the candied peel into short pieces about 1cm thick and the length of a matchstick. Dip each piece of peel into the chocolate, place on a piece of baking parchment and leave in a cool place until the chocolate has set.

Pour the double cream into the fruit juice and sherry, then slowly whip until the syllabub starts to thicken. It is best to whip slowly, watching the consistency of the syllabub constantly. As it starts to thicken, the cream will feel heavy on the whisk and you should stop as soon as the syllabub is thick enough to sit in soft, gentle waves. If you whisk until it will stand in peaks you have gone too far. Keep chilled in the refrigerator.

Remove the peel from the remaining blood oranges with a very sharp knife, taking every shred of white pith with you. Slice the fruit thinly. Spoon the chilled syllabub into bowls, then add the slices of blood orange and the chocolate-dipped candied peel.

RICE, MILK, FIG JAM

The humble sweetness of baked rice. The syrupy luxury of fig preserve.

Serves 4

green cardamom pods 6
milk 1 litre
bay leaves 2
a vanilla pod

butter 30g
pudding rice (short-grain) 80g
caster sugar 3–4 tablespoons
fig jam/preserve 8 tablespoons

Set the oven at 140°C/Gas 1.

Crack the green cardamoms, extract the tiny brown-black seeds, then crush them to a powder. A pestle and mortar will work if you don't have a spice mill. Pour the milk into a saucepan. Add the ground cardamom, bay leaves, vanilla pod and butter and bring to the boil.

Put the pudding rice and caster sugar in a baking dish, then, as soon as the milk boils (watch carefully), pour the hot milk over the rice and stir till the sugar has dissolved.

Slide the dish into the preheated oven and bake for two hours until the rice is soft and creamy and the skin is pale gold. Serve with spoonfuls of soft fig jam.

• No effort involved here. Just a little patience.
• If fig jam proves tricky to track down, slice six small fresh figs in half and put them in a saucepan with four tablespoons each of granulated or caster sugar and water. Bring to the boil, then lower the heat and simmer till the sugar and water have turned to a syrup and the figs are soft and on the point of collapse.

Pudding   285

Index

fig jam:
  rice, milk, fig jam 282
figs:
  mushrooms, orange, breadcrumbs 138
filo pastry, cheese, greens 146
fregola, greens, Pecorino 200
fritters:
  butternut, feta, eggs 24
  polenta, thyme, Taleggio 226

G
garam masala:
  aubergine, tomato, coconut milk 196
  beetroot, lentils, garam masala 166
garlic:
  parsnips, smoked garlic, feta 104
  pearl barley, kale, goat's cheese 220
  polenta, garlic, mushrooms 222-4
  potatoes, Brussels sprouts 106
  potatoes, tomatoes, horseradish 112
ginger:
  apples, ginger cake, custard 246
  aubergines, ginger, tamarind 14-16
  ginger cake, cardamom, maple syrup 264
gnocchi, peas, egg yolk 202
goat's curd:
  beetroot, apple, goat's curd 48
  parsnips, shallots, goat's curd 102
  pearl barley, kale, goat's cheese 220
golden sultanas:
  apples, pears, rye 250
  chocolate, dried fruit, pistachios 256
  Jerusalem artichokes, pistachios,
    grapes 32
  oats, dried mulberries, date syrup 6
grapes:
  cheese, thyme, grapes 88
  Jerusalem artichokes, pistachios,
    grapes 32

H
haricot beans:
  artichokes, beans, green olives 12
  haricot beans, rice, onions 204
  noodles, lentils, soured cream 182-4
harissa paste:
  sweet potato, puff pastry 158-60
hazelnuts:
  chocolate, dried fruit, pistachios 256
  hazelnuts, blood oranges, ricotta 266-8
  mushrooms, orange, breadcrumbs 138
honey:
  hazelnuts, blood oranges, ricotta 266-8
horseradish:
  beetroot, sauerkraut, dill 168
  celeriac, horseradish, pumpernickel 174
  potatoes, tomatoes, horseradish 112
hummus:
  mushrooms, hummus, herbs 34
  pumpkin, chickpeas, rosemary 114-16

I
ice cream, chocolate, sherry 270

J
jalapeño chillies:
  sweet potato, jalapeños, beans 124
Jerusalem artichokes:
  artichokes, winter roots, smoked
    salt 68-70
  Jerusalem artichokes, pistachios,
    grapes 32

K
kale:
  black-eyed beans, rosemary, kale 170
  pearl barley, kale, goat's cheese 220

L
labneh:
  mograbia, tomatoes, labneh 212

O

oats:
  apples, oats, chocolate 248
  oats, dried mulberries, date syrup 6
  winter porridge 3
olives:
  artichokes, beans, green olives 12
omelettes:
  eggs, edamame, bean sprouts 26–8
onions:
  a brown vegetable stock 4
  cauliflower, onions, bay 82
  Cheddar, cider, mustard 176
  haricot beans, rice, onions 204
  lentils, swede, paprika 206–8
  mograbia, tomatoes, labneh 212
  noodles, lentils, soured cream 182–4
  onions, Taleggio, cream 100
  parsnips, cashews, spices 218
  parsnips, smoked garlic, feta 104
  potatoes, sweet potatoes, cream 108
  pumpkin, onions, rosemary 38
  (see also shallots)
oranges:
  beetroot, blood orange, watercress 136
  chickpeas, radicchio, butter beans 90
  chocolate, orange zest, candied peel 260
  fennel, cream, pine kernels 92
  hazelnuts, blood oranges, ricotta 266–8
  mushrooms, orange, breadcrumbs 138
  papaya, passion fruit, mint 274
  rice, lemon, lime 230
  sherry, blood oranges, cream 280
orecchiette, cauliflower, cheese 216
orzo, smoked mozzarella, thyme 186
oyster mushrooms:
  mushrooms, hummus, herbs 34
  mushrooms, spinach, rice 214
  tagliatelle, dill, mushrooms 236

P

pak choi:
  eggs, edamame, bean sprouts 26–8
  rice, broccoli, pak choi 228
paneer:
  carrots, spices, paneer 80
  papaya, passion fruit, mint 274
pappardelle:
  lentils, swede, paprika 206–8
  parsley, Parmesan, eggs 36
parsnips:
  artichokes, winter roots, smoked
    salt 68–70
  leeks, parsnips, pastry 148
  parsnips, cashews, spices 218
  parsnips, shallots, goat's curd 102
  parsnips, smoked garlic, feta 104
passion fruit:
  papaya, passion fruit, mint 274
pasta:
  gnocchi, peas, egg yolk 202
  lentils, swede, paprika 206–8
  orecchiette, cauliflower, cheese 216
  orzo, smoked mozzarella, thyme 186
  tagliatelle, dill, mushrooms 236
pastries:
  apples, cinnamon, puff pastry 244
  filo pastry, cheese, greens 146
  sweet potato, puff pastry 158–60
  (see also pies; tarts)
peanuts:
  mushrooms, spinach, rice 214
pearl barley, kale, goat's cheese 220
pears:
  apples, pears, rye 250
  pears, red wine, Pecorino 276
  red cabbage, carrots, smoked almonds 140
peas:
  fennel, peas, halloumi 30
  gnocchi, peas, egg yolk 202

Acknowledgements

A heartfelt thank you.

The messages by email and social media have been extraordinary. Heart-warming, generous, encouraging, each one has been gratefully received. The response to the first volume of *Greenfeast*, and the idea that 'yes, this is how we want to eat now' has been life-affirming. That there are so many who wish to eat food that is predominantly vegetarian and plant-based rather than meat-led or strictly vegetarian, has been welcome news. I want to thank everyone who has been in touch about the first volume of *Greenfeast* and to say that it is your enthusiasm that has led to this, the second volume being finished on time.

That the books are here at all is due to the constant support and patience of everyone at 4th Estate, especially Louise Haines, my editor for as long as I have been writing. Thanks too, to everyone at The Soho Agency and United Agents.

The recipes, as always, have been a collaboration with James Thompson and it was once again a joy working with photographer Jonathan Lovekin, with whom I have worked every week for over thirty years.

The books are designed not just to look pretty on a shelf – but to be used, day in day out, in the kitchen. Compact, tactile, easy to use. Books that look better with use. (I love the way a book settles in with use and opens flat at the most cooked-from pages.) The fact that they are everything I hoped them to be is due to Tom Kemp,

Julian Humphries, David Pearson, Chris Gurney, Gary Simpson, Jack Smyth and to everyone at GS Typesetting and Neographia.

While many recipes are new to this collection, several started their life in the pages of the *Observer* magazine, and I thank Allan Jenkins, Harriet Green, Martin Love, Gareth Grundy and the entire team for their continuing support and encouragement.

Now, here's the thing. There is no point in anyone putting pen to paper, or more accurately, fingers to a keyboard, no reason for cooking and illustrating and printing and publishing, if there is no-one to send those words and pictures out into the world. I want to shout a sincere thank you to the booksellers of the world who continue to fill our lives with so much joy. The booksellers who look after our books, who embrace new volumes whilst keeping older, much-loved titles on their shelves, and especially those who recommend books to those of us who walk through their doors, hungry with anticipation. There is almost nowhere I would rather be than within the walls of a good bookshop.

Thank you one and all.

Nigel Slater, London, September 2019

A note on the brushstrokes

Many of us are placed on this planet in such a way that we experience all four seasons. For our cultures they constitute a deeply ingrained metaphor for growth and change, reinforced throughout our lives by seeing almost every plant (and certainly the ones in this book) attempting to reproduce itself annually, seasonally, in lockstep with the Earth's tilt and orbit. We even map the seasons on to our lives: as someone who has reached 'autumn' I finally understand what older people have been telling me all my life and also the impossibility of any younger person understanding a word of what was said!

Spring and summer represent a forever forgetful beginning. The cold, barren, enclosedness of winter is gleefully abandoned the moment the sun begins to rise higher and sooner in the sky. There is a looking-forward to the growth, energy and abundance which are about to follow. The later seasons, however, tend to be soaked in a melancholy regret for all that exuberance and naivety; redeemed, certainly, by the warm, homely comfort of hibernation but also by the knowledge that longer, warmer days will come round again.

However, when it's one's life being measured by the seasons, there is no going round again. There is no undoing of the awful mistakes, no reliving of the youthful joy. There is a much deeper appreciation of what life is; a much bolder outlook on the importance of error, of hesitancy, of not having all the answers, of being more and more open to possibility.

The brushstrokes in this book are, again, not pictures of things. Each one is a record of a thought or feeling or simply a movement manifested through my use of a square-edged brush for a few concentrated seconds. They are informed by what came before, but they have a knowingness, a memory about them which was not possible for those in the first book. I've let them be a little more idiosyncratic and sometimes not so sure of their path.

The square tip of the brush is like a large version of the end of the quill pen used by medieval scribes. The shapes it makes are mostly determined by the directions and angles I choose to push the brush around, rather than the pressure which informs the marks made with a pointed brush. Each brushstroke is ultimately derived from a letterform practised over and over again to the point of abstraction and together they define a kind of illegible but meaningful text.

It's a small relief to me to be able to record my existence in this very intimate and precise way for the very brief time the brush is in contact with the surface. These strokes are like deciduous leaves providing temporary evidence of the tree from which they fell.

Tom Kemp

A note on the type

The text is typeset in Portrait, designed by Berton Hasebe in 2013. Portrait is a modern-day interpretation of the early Renaissance typefaces, first made popular in Paris and cut by Maître Constantin. His influence can be seen in the work of the famed punch cutters of the next generation, Claude Garamond and Robert Granjon, whose work is so familiar to modern day readers.

The cover type is set in Brunel, an English modern designed by Paul Barnes and Christian Schwartz in 2008. Brunel in turn was based on the first moderns issued by the Caslon foundry in 1796. The name is derived from the Anglo-French engineers Sir Marc and Isambard Kingdom Brunel.